DATA FUSION AND SENSOR MANAGEMENT:
A Decentralized Information-Theoretic Approach

ELLIS HORWOOD SERIES IN ELECTRICAL AND ELECTRONIC ENGINEERING

Series Editors: D. R. SLOGGETT, Business Development Director, Earth Observated Sciences Limited, Farnham, Surrey and Professor E. H. MAMDANI, Department of Electronic Engineering, Queen Mary and Westfield College, University of London

P.R. Adby	APPLIED CIRCUIT THEORY: Matrix and Computer Methods
J. Beynon	PRINCIPLES OF ELECTRONICS: A User-Friendly Approach
R.L. Brewster	TELECOMMUNICATIONS TECHNOLOGY
R.L. Brewster	COMMUNICATION SYSTEMS AND COMPUTER NETWORKS
S.J. Cahill	DIGITAL AND MICROPROCESSOR ENGINEERING: Second Edition
C. Dislis, J. Dick, I.D. Dear, A. P. Ambler	TEST ECONOMICS AND DESIGN FOR TESTABILITY
J. Jordan, P. Bishop & B. Kiani	CORRELATION-BASED MEASUREMENT SYSTEMS
A. Klos & P. Aitchison	NETWORK SOLUTION: theory and practice
J.Manyika & H. Durrant-Whyte	DATA FUSION AND SENSOR MANAGEMENT
P.G. McLaren	ELEMENTARY ELECTRIC POWER AND MACHINES
P. Naish & P. Bishop	DESIGNING ASICS
J.R. Oswald	DIACRITICAL ANALYSIS OF SYSTEMS: A Treatise on Information Theory
J. Richardson & G. Reader	ANALOGUE ELECTRONICS CIRCUIT ANALYSIS
J. Richardson & G. Reader	DIGITAL ELECTRONICS CIRCUIT ANALYSIS
J. Richardson & G. Reader	ELECTRICAL CIRCUIT ANALYSIS
P. Sinha	MICROPROCESSORS FOR ENGINEERS: Interfacing for Real Time Applications
J.N. Slater	CABLE TELEVISION TECHNOLOGY
J.N. Slater & L.A. Trinogga	SATELLITE BROADCASTING SYSTEMS: Planning and Design
R. Smith, E.H. Mamdani, S. Callaghan	THE MANAGEMENT OF TELECOMMUNICATIONS NETWORKS
E. Thornton	ELECTRICAL INTERFERENCE AND PROTECTION
L.A. Trinogga, K.Z. Guo & I.C. Hunter	PRACTICAL MICROSTRIP CIRCUIT DESIGN
J.E. Whitehouse	PRINCIPLES OF NETWORK ANALYSIS
Wen Xun Zhang	ENGINEERING ELECTROMAGNETISM: Functional Methods
A M S. Zalzala & A S Morris	NEURAL NETWORKS IN ROBOTIC CONTROL THEORY & APPLICATION

Of related interest

ELECTRONIC AND COMMUNICATION ENGINEERING

R.L. Brewster	TELECOMMUNICATIONS TECHNOLOGY
R.L. Brewster	COMMUNICATIONS AND COMPUTER NETWORKS
J.N. Slater	CABLE TELEVISION TECHNOLOGY

ELLIS HORWOOD SERIES IN DIGITAL AND SIGNAL PROCESSING

Series Editor: D.R. SLOGGETT, Technical Director, Marcol Group Ltd, Woking, Surrey

G.D. Bergman	ELECTRONIC ARCHITECTURES FOR DIGITAL PROCESSING: Software/Hardware Balance in Real-Time Systems
Z. Hussain	DIGITAL IMAGE PROCESSING
R. Lewis	PRACTICAL DIGITAL IMAGE PROCESSING
S. Lawson & A. Mirzai	WAVE DIGITAL FILTERS

DATA FUSION AND SENSOR MANAGEMENT:
A Decentralized Information-Theoretic Approach

J. Manyika and H. Durrant-Whyte
Department of Engineering Science,
University of Oxford, UK

ELLIS HORWOOD
NEW YORK LONDON TORONTO SYDNEY TOKYO SINGAPORE

First published 1994 by
Ellis Horwood Limited
Market Cross House, Cooper Street
Chichester
West Sussex, PO19 1EB
A division of
Simon & Schuster International Group

© Ellis Horwood Limited 1994

All rights reserved. No part of this publication may be reproduced, stored in a retrieval system, or transmitted, in any form, or by any means, electronic, mechanical, photocopying, recording or otherwise, without prior permission, in writing, from the publisher

Printed and bound in Great Britain by
Bookcraft, Midsomer Norton

Library of Congress Cataloging-in-Publication Data

Available from the publisher

British Library Cataloguing in Publication Data

A catalogue record for this book is available from the British Library

ISBN   0-13-303132-2   (hbk)

1 2 3 4 5   97 96 95 94 93

To my Mother and to the memory of my Father and all that he stood for.

James Manyika

Contents

Preface xiii

Acknowledgements xvi

List of Figures xviii

Nomenclature xvi

1 Introduction and Overview 1
  1.1 Sensing and Multi-Sensor Systems 1
  1.2 Data Fusion and Sensor Management 4
    1.2.1 A Taxonomy of Issues 6
    1.2.2 Background 7
  1.3 An Information-Theoretic Approach 12
  1.4 Application and Practical Demonstration 14
  1.5 A Guided Tour of the Book 17

2 A Probabilistic Model for Managed Data Fusion 20
  2.1 Observations, Inference and Estimation 21
    2.1.1 A Geometrical Representation of Uncertainty . . . 22
    2.1.2 A Bayesian Paradigm 23
    2.1.3 Classical Inference and Estimation 25
  2.2 Combining Probabilistic Information 28
    2.2.1 Pooling Information 29
    2.2.2 Independent Likelihood Pool 32
    2.2.3 Remarks 33
  2.3 Decisions and Actions 35
    2.3.1 Decisions and the Single Bayesian 35
    2.3.2 Decisions involving Multiple Information Sources . 36
    2.3.3 Utility Theory 40

	2.4	The Nature and Role of Information	42
		2.4.1 Measures of Probabilistic Information	42
		2.4.2 Probabilistic Information Update	45
		2.4.3 Initialization, Priors and Robustness	46
	2.5	Summary	47
	2.6	Bibliographical Note	48
3	**Architectures and Algorithms**		**50**
	3.1	A Progression of Data Fusion Architectures	51
	3.2	Architectures and Information Update	56
		3.2.1 A Generic Single Sensor Architecture	56
		3.2.2 Centralized and Hierarchical Architectures	56
		3.2.3 Distributed and Decentralized Architectures	60
		3.2.4 Remarks	62
	3.3	Multi-Sensor Continuous State Estimation	63
		3.3.1 The Information Filter	64
		3.3.2 Extension to Hierarchical Estimation	72
		3.3.3 Decentralized Information Filter	74
		3.3.4 On Initialization, Consistency and Optimality	80
	3.4	Classifying Discrete States and Objects	82
		3.4.1 A Bayesian Approach to Classification	82
		3.4.2 Hierarchical Classification	83
		3.4.3 Decentralized Classification	84
		3.4.4 On time varying attributes and different frames of discernment	86
	3.5	A Note on Data Association and Validation	87
	3.6	Bibliographical Note	88
4	**Data Fusion Management**		**90**
	4.1	Elements of a Normative Formulation	91
		4.1.1 Management Imperative, Actions and Outcomes	91
		4.1.2 The Utility of Information	94
	4.2	Data Fusion Information	98
		4.2.1 Information Filter Metrics	98
		4.2.2 Metrics for Bayesian Classification	101
	4.3	Towards a Decentralized Sensor Management Solution	102
		4.3.1 Information Available to a Sensor Node	102
		4.3.2 The Probabilistic Implications of Actions	104
		4.3.3 Comparable and Non-comparable Utility Solutions	108
		4.3.4 Formulation for Fusion Algorithms	114

Contents

    4.4 Realizing Decentralized Management 115
         4.4.1 Computation, Communication and Bargaining . . . 115
         4.4.2 An Iterative (Bargaining) Algorithm 121
    4.5 A Necessary Discussion 125
         4.5.1 On Rationality and Optimality 125
         4.5.2 Is it worth the bother? 127
         4.5.3 Coupled Management of Data Fusion Algorithms . 128
         4.5.4 Summary 130
    4.6 Bibliographical Note 130

5 **A Sensor Model**   **132**
    5.1 Sonar in Mobile Robotics 133
    5.2 Understanding the Sensor 134
         5.2.1 Sonar Measurements and Interpretation 134
         5.2.2 Sensing Limitations 136
         5.2.3 Differential Sonar Model 138
    5.3 A Tracking Sonar Sensor 140
    5.4 Sensor Uncertainty and Probabilistic Model 144
         5.4.1 Differential Uncertainty 144
         5.4.2 Tracking Sonar Uncertainty 147
         5.4.3 Use of Spectral Estimation Techniques 151
         5.4.4 Analysing Measurement Uncertainty 154
         5.4.5 Measurement Probabilistic Model 158
    5.5 Discussion 160
         5.5.1 Summary of Performance and Limitations 160
         5.5.2 A Modular Data Fusion Sensor 162
    5.6 Bibliographical Note 163

6 **Data Fusion for Robot Navigation**   **165**
    6.1 Mobile Robot Navigation 165
    6.2 Estimation of Location 167
         6.2.1 Sensor-based Feature Descriptions 167
         6.2.2 Localization Algorithm 169
         6.2.3 Implementation 171
         6.2.4 Results from estimating the location of a sensor
                platform 174
    6.3 A Decentralized Localization Scheme 177
         6.3.1 Algorithm and Implementation 177
         6.3.2 Decentralized Results and Performance 179
    6.4 OxNav Vehicle Navigation: A Discussion 185

		6.5	Modular Feature Classification 190
			6.5.1 Using Displacement to Classify Features 190
			6.5.2 Algorithm and Implementation 192
			6.5.3 Some Classification Results 194
		6.6	Concluding Remarks and Other Issues 198
		6.7	Bibliographical Note 200

7 Sensor Management Demonstrations   201

    7.1 Sensor Management on a Navigating Robot 202
    7.2 Quantities for Sensor Management 203
        7.2.1 Variations in State Estimation Metrics 203
        7.2.2 Variations in Classification Uncertainty Metrics . . 206
        7.2.3 Discussion 210
    7.3 Sensor-Feature Assignments 211
        7.3.1 Demonstration 1 211
        7.3.2 Demonstration 2 214
        7.3.3 Demonstration 3 214
    7.4 Sensor Hand-off and Cueing 217
        7.4.1 Demonstration 4 217
        7.4.2 Demonstration 5 221
    7.5 Summary 221
    7.6 Bibliographical Note 222

8 Towards General and Robust Managed Data Fusion   223

    8.1 Review of Significance and Limitations 223
    8.2 On General Methods and Paradigms 225
        8.2.1 Developing General Normative Management 225
        8.2.2 Generally Applicable Paradigms 227
    8.3 Robust and Fully Functional Systems 229
        8.3.1 Robustness 229
        8.3.2 Mobile Robot Navigation 231

Appendices   232

A On Entropy and Information   232

    A.1 Entropy of a vector distribution given covariance 232
    A.2 Non-informative priors and Entropy 233
    A.3 On the relationship between Fisher information and entropy 235

Contents

B Differential Sonar Details   **237**
- B.1 Sonar physical model 237
- B.2 Differential feature model 239
- B.3 Tracking Sonar performance and limitations 243
- B.4 Hardware 246
  - B.4.1 Generic decentralized sensor node architecture . . 246
  - B.4.2 Differential sonar 247

Bibliography   **249**

Index   **263**

Where is the life we have lost in living?
Where is the wisdom we have lost in knowledge?
Where is the knowledge we have lost in information?

- T.S. Eliot (The Rock 1934)

Preface

The aim of this monograph is to present a framework for addressing multi-sensor data fusion and sensor management in general, and in decentralized systems in particular. And, subsequently, to apply and demonstrate the resulting algorithms and methods in the context of an extant research problem in mobile robotics.

The theory and application of multi-sensor systems has been approached from widely differing perspectives most of which betray the applicators' theoretical backgrounds and interests. Hence, computer scientists have taken approaches founded on artificial intelligence techniques and the use of knowledge bases, electrical engineers have used statistical estimation and signal processing techniques, while mathematicians and statisticians have used interval calculus, Bayesian paradigms etc. All these, seemingly, diverse approaches are aimed explicitly at addressing the fundamental problem of how to combine, in the best possible manner, diverse and uncertain sensor measurements and other information available in a multi-sensor system. The ultimate aim is to enable the system to estimate or make inferences concerning the state of nature (e.g. the state of a robot's environment). Such combination of information and subsequent inference of the state is the *Data Fusion* process. With data fusion defined in this way, an additional matter arises, which is the question of how best to manage, coordinate and organize sensing resources in a multi-sensor system. This defines the *Sensor Management* problem. To complicate matters further, multi-sensor systems now assume different architectural forms, ranging from centralized and hierarchical systems, to the ascendant distributed and decentralized systems and also various hybrid systems. The ascendancy of distributed and decentralized systems is due to their significant inherent operational advantages and also to the current trends towards modular and autonomous systems.

To date, data fusion and sensor management have largely been dealt with separately and primarily for centralized and hierarchical systems. Although research has been done recently in distributed and decentralized systems, very little of it has addressed both data fusion and sensor management. When applied to distributed and decentralized systems, the oft ad hoc methods used in other multi-sensor systems simply become unjustifiable, while some of the well-founded approaches cannot be extended easily to cope with the added complexities of these systems. Having had considerable first hand theoretical and practical experience working on the various issues encountered in multi-sensing, it became increasingly apparent to us that many, if not all, of the issues arising in multi-sensing can be addressed within a single consistent framework.

Therefore, this monograph is our personal statement on the development of precisely such a single and consistent framework in general and in decentralized systems in particular. Our approach is based on considering information and its gain as the *raison d'être* of multi-sensor systems. The concept of information and its gain can be formalized to give a probabilistic information update paradigm from which architectures and algorithms for data fusion can be developed *directly*. The information update paradigm also leads to an intuitive method of addressing sensor management; when presented with several sensing alternatives, sensor management addresses the question of how to make decisions leading to the best sensing configuration or actions. Stated in this way, sensor management reduces to a problem in Decision Theory for which an intuitive basis for making decisions is a consideration of the value of the sensing information obtained. This leads to the development of a *normative* method for sensor management which makes use of information metrics as the expected utility.

Underlying such an information-theoretic approach is the need to understand the nature of the information obtained by each sensor. This requires a good sensor model detailing each sensor's physical operation and the phenomenological nature of its measurements *vis-à-vis* the probabilistic information these measurements provide. Therefore, in order to present practical techniques for developing sensor models, we describe a *Tracking Sonar* sensor with a focus of attention capability. As already stated, implicit in a sensor management problem is the existence of either several sensing strategies such as those provided by agile or multi-mode sensors, or several alternative configurations for the multi-sensor system. In this regard, the sensor that we discuss, as a result of its mode of operation, also serves the purpose of facilitating a non-hypothetical discussion on the practical management of sensors. Conducting a sensor model discussion in the context of a particular sensing modality i.e. sonar, allows us to present the model in some detail which provides considerable insight into the practical development of sensor models. However, this is at the expense of perhaps a much

Preface

broader discussion across several sensing modalities. Nevertheless, the techniques presented can be extended to other sensors.

We are firm believers that real applications and compelling demonstrations are the hallmarks of good engineering theories, not least given the skepticism of most engineers including ourselves. And so, in order to lend credence to our theoretical work and also to provide non-trivial examples of how the methods can be used, we describe in some detail a real application to mobile robot navigation. More specifically, we consider mobile robot localization and map-building. The implementation described is part of the work we have done on the OxNav[1] project, which aims to develop a mobile vehicle capable of navigating in indoor environments. For this purpose, we have designed and purpose-built *Joey*, a vehicle which makes use of modular autonomous sensing and kinematic units. The modular philosophy behind the OxNav project makes *Joey* an ideal platform for demonstrating decentralized architectures, algorithms and management techniques. The navigation results shown are thus not simulations but results obtained during actual runs. We show firstly how, by managing data fusion, localization is speeded up and secondly, how the allocation, hand-off and cueing of sensors in a multi-sensor system can be implemented decentrally and autonomously.

It is important to emphasize that while vehicle navigation is the only application described in detail, the methods are applicable to a host of other multi-sensing problems. Indeed the abstraction of the overall problem to one of managing and extracting information from a multi-source system under conditions of uncertainty makes our approach amenable to a variety of quite diverse applications, some outside the bounds of robotics. However, focussing on one application enables us to present actual implementations in some detail.

Since what we present is essentially intended as a framework, we have not shown exhaustively how *every* problem in multi-sensing can be addressed. We hope that what we have presented gives some indication as to how other, more specific, data fusion and sensor management problems can be addressed. While we have attempted to make the text as accessible as possible, we have assumed a basic understanding of; probability and estimation theory, linear and non-linear systems, an

---

[1]The **Ox**ford **Nav**igator project was funded by an SERC-ACME grant GR 38375. The success of the original OxNav vehicle *Joey* has resulted in other similar vehicles , i.e. *Elvis* and *Sharon*.

appreciation of sensors and sensing. Where we have felt it unnecessary to develop certain concepts from first principles we have made reference to some of the most popular texts and papers covering the respective topics. Since the material in this monograph is presented somewhat *parti pris*, we have provided an extensive bibliography which includes references to different and contrasting approaches to our own.

The text is intended to be read through as a monograph. However, readers whose primary interest is in the theory and algorithms for data fusion and sensor management will perhaps find it convenient to concentrate on Chapters 1 to 4 and only make reference to later chapters. Chapter 2 will be found particularly useful for readers interested in the theoretical foundations for probabilistic methods used in data fusion. Chapters 5 may be skipped in a cursory reading as it gives a detailed sensor model for sonar. Chapter 6 and 7 are implementation chapters which, application-oriented, data fusion practioners will find particularly useful. Readers whose primary interest is in mobile robot navigation will find it useful to concentrate on Chapters 5 to 7 whilst only making reference to Chapters 3 and 4. A guided tour through the book is provided at the end of Chapter 1.

Acknowledgements

The work outlined in this text was carried out almost in its entirety in the Robotics Research Group at Oxford University. We are indebted to Prof. Mike Brady for his keen interest in the work. We thank colleagues on the OxNav project, Tom Burke, Michael Stevens and Ian Treherne for software and hardware support, and without whom there would be no *Joey*. Our gratitude also goes to other colleagues, Stewart Grime, Peter Ho, Billur Barshan for their many useful comments, constructive criticism and support. We also thank Dr. John Leonard (MIT) and Dr. Bobby Rao (Berkeley), who laid the foundations for much of this work, for the numerous fruitful discussions and helpful comments. Thanks are due to Dr. Paul Schenker (NASA Jet Propulsion Laboratory), and Dr. Alec Cameron (Philips Research) from whom, through visits and discussions, we benefited immensely. We are also indebted to the many helpful comments and suggestions by Dr. Ron Daniel and Dr. John Hallam (Edinburgh). We are very grateful to our "amanuensis" Sarah Ladipo for assistance in proof-reading part of the manuscript. We also

Acknowledgements

thank our publisher Ellis Horwood for their assistance. The bulk of this research was funded through SERC-ACME grant GR 38375.

James Manyika and Hugh Durrant-Whyte

I must thank the many members of my three colleges, Keble, St Hugh's and Balliol for making my Oxford experience extremely varied and enjoyable. Thanks go to my many friends at Oxford especially, David, Liz, Michele, Phil and Vel for their companionship, tolerance and understanding. I would like to thank my mother and Sarah for their ever present love and support. Most importantly, I wish to pay tribute to my father James M.D. Manyika who died as I completed the research for this monograph. I miss him and thank him for everything. Finally, in order to avoid the worst hell of all in Dante's Inferno which is reserved for those who do not give due respect to their benefactors, I would like to thank the Rhodes Trustees who through the generosity of a Rhodes Scholarship funded my research at Oxford. In addition, thanks are due to St Hugh's College for the award of a Smith-Rippon Senior Scholarship and to the Fellows of Balliol College for election to a Research Fellowship.

James Manyika,
Balliol College, Oxford University.
Michaelmas 1993

List of Figures

1.1 Sensing. From a sensor device measurement to perceptual information. 2
1.2 Sensor data fusion. 5
1.3 A taxonomy of issues in Multi-Sensor Data Fusion. 6
1.4 A Road Map. Each *Ch* refers to the appropriate chapter. . 19

2.1 Uncertainty ellipsoid for a 2-dimensional state. 23
2.2 Linear Opinion Pool. 30
2.3 Independent Opinion Pool. 31
2.4 Independent Likelihood Pool. 33
2.5 "Super" Bayesian approach to management decision-making. 38
2.6 Multi-Bayesian approach to management decision-making. 38

3.1 Centralized Architectures. 51
3.2 Hierarchical Architecture. 52
3.3 Decentralized Architectures. (a) fully connected (b) non-fully connected. 54
3.4 Single sensor information update. 56
3.5 Centralized information update for architecture communicating local posterior information to central processor. . 58
3.6 Centralized information update for architecture with sensor communicating actual observations. 59
3.7 Centralized information update for architecture communicating likelihoods to central processor. 59
3.8 Decentralized information update at each sensor node. . . 62

4.1 Mapping from action space through probabilistic results set to inferred state $\hat{x}$. 92
4.2 Elements of a normative approach to sensor management. 93

List of Figures

4.3 Probability distributions arising from actions a_1 and a_2 in **Example 1**. 96
4.4 The decentralized sensor system of **Example 2**. 105
4.5 Some of the alternate sensing strategies implied by elements of the action set $\mathcal{A}$ corresponding to Figure 4.4. . . . 106
4.6 Computing the expected utilities for every sensor in the system at sensor i. 118
4.7 Communicating expected utilities for sensor i locally and then communicating to other sensors. 119
4.8 1st and 2nd action preferences for **Example 2**. 121
4.9 An iterative algorithm for making decentralized management decisions. 123
4.10 Depicting the time variations of utility functions in relation to computing more iterations of the bargaining algorithm. 127

5.1 (a) Illustrates angle of inclination to the direction of propagation α and (b) parameters in the RCD model. 135
5.2 Sector scan showing (a) range data (b) RCDs extracted. . . 136
5.3 Illustrating the differential principle for a positive value of α. 139
5.4 An extracted plane RCD and its associated differential. . . 139
5.5 Information obtained from a return within the bounds of an RCD. 141
5.6 Tracking Sonar Algorithm. 142
5.7 Illustrating motion of a tracked feature in sensor egocentric coordinates 143
5.8 Plots showing multiple (5) differentials within the width of the RCD corresponding to each of the four target types. 148
5.9 Differentials measured at a fixed orientation within the bounds of an RCD, highlighting the uncertainty associated with the differential. 149
5.10 Plots (a) and (b) show the measured and the estimated differentials. 149
5.11 Results showing the measured and estimated differentials while tracking a feature. 150
5.12 Plots showing the variation in bearing θ over time with no relative motion between the feature and the sensor for various differential estimator noise models σ_δ^2. 155

5.13 Autocorrelation estimates for the orientation during tracking of a stationary feature. 156
5.14 Power spectra using a Bartlett window $m = 512$ for the orientation during tracking for a stationary feature. 157
5.15 A completed modular Tracking Sonar sensing node. 163

6.1 Sensor-based feature descriptions in a global coordinate system. 168
6.2 Localization algorithm. 173
6.3 Localization estimates for a stationary sensor platform using various observation noise models. 174
6.4 Localization estimates for (a) motion in y-axis and (b) motion in x-axis. 175
6.5 Location estimates during motion between two known locations. 176
6.6 Tracking Sonar configuration for the OxNav vehicle. . . . 178
6.7 Decentralized location estimates (showing x-axis only) for a stationary vehicle using 2 Tracking Sonars. 181
6.8 Decentralized location estimates for a vehicle which is moving in the x-axis. 181
6.9 Comparing estimates with hand measured values while localizing on the same feature at different ranges. Here vehicle motion is not smooth. 182
6.10 Decentralized location estimates with smooth vehicle motion while sensors track the same feature. 183
6.11 Decentralized location estimates with smooth vehicle motion while sensors track different features. 183
6.12 Decentralized estimates for motion incorporating 2 turns based on a trajectory as defined by a spline for turning a corner in a continuous motion. 184
6.13 The OxNav Vehicle *Joey*. 186
6.14 Joey navigating down a corridor. 187
6.15 Joey navigating in a cluttered environment. 188
6.16 A classification model for observations from two positions (x_1, y_1) and (x_2, y_2). 190
6.17 Observation model for an arbitrary feature illustrating the parameters used for classification. 191
6.18 Variation with classification model for a plane. 194
6.19 Variation with classification model and value of minimum displacement for a plane. 196

List of Figures

6.20 Variation with classification model and value of minimum displacement for a plane. 197

7.1 Variation of information quantities with range for various observation models. 205
7.2 Information quantities for motion past a tracked feature. . 205
7.3 Variation of classification information with parameter model. 207
7.4 Variation of classification information with displacement step ξ. 208
7.5 More variation of classification information with displacement step ξ. 209
7.6 Cumulative classification observed information. 211
7.7 Variation of information quantities with two sensor-feature assignments while the vehicle is rotating counter-clockwise. 212
7.8 Variation of information quantities with two alternate sensor-feature assignments while the vehicle moves as indicated. 215
7.9 Various sensing strategies for a vehicle with two sensors and able to track two features. 216
7.10 Total global information corresponding to each of the actions in Figure 7.9. 216
7.11 Partial and observation information used for hand-off while vehicle motion is in a straight line. 218
7.12 Partial and observation information used for hand-off as vehicle moves back and forth. 218
7.13 Information values for two sensors running the decentralized classification algorithm while tracking the same feature. 219
7.14 Partial information at each sensor for two sensors classifying the same feature. 220

B.1 (a) The observed time waveform (b) The impulse response of the receiver. 238
B.2 The effect of varying frequency. 239
B.3 Differential sonar model for various features. 240
B.4 Generic Transputer-based decentralized sensor node hardware. 247

Nomenclature

Notation

$(\cdot)^T$	Matrix transpose
$(\cdot)^{-1}$	Matrix inverse
$\bar{(\cdot)}$	Mean of $(\cdot)$
$\hat{(\cdot)}$	Estimate of $(\cdot)$
$\tilde{(\cdot)}$	Partial estimate of $(\cdot)$
$E\{\cdot\}$	Expected value of $\{\cdot\}$
$a \triangleq b$	a is defined as b
$\{(\cdot)\}$	Set whose elements are $(\cdot)$
$<\cdots>$	Ordered set
$(\cdot)(k)$	$(\cdot)$ at time k
$(\cdot)(k \mid l)$	$(\cdot)$ at time k given l

General Symbols

The following symbols are used consistently throughout this book. However, in some chapters there are some locally defined symbols whose scope is limited to those chapters only.

Symbol	Interpretation
k	Time step
$\mathcal{P}$	Probabilistic set
$p(\cdot)$	Probability distribution where $p \in \mathcal{P}$
λ	Likelihood function
Λ	Likelihood vector
$\mathcal{X}$	State space
$\mathbf{x}$	State vector $\mathbf{x} \in \mathcal{X}$
X_l	lth element in a discrete state vector
$\mathbf{P}$	State covariance matrix
$\mathbf{y}$	Information state vector
$\mathcal{Z}$	Observation space
$\mathbf{z}$	Observation vector
$\mathbf{Z}^k$	Set of observation vectors $\mathbf{z}$ up to time k
$\mathbf{Z}_j^k$	Set of observation vectors up to k for sensor j

Nomenclature

General Symbols(cont)

Symbol	Interpretation
$\mathcal{V}$	Observation noise space
v	Observation noise vector
R	Observation noise covariance
$\mathcal{M}$	General observation model
$\{M\}$	Set of observable parameters
M_l	lth observable parameter
H	Observation matrix
h[·]	Non-linear observation model
w	State transition noise
Q	State transition noise covariance
F	State transition matrix
f[·]	Non-linear state transition model
u	Control input
$\mathcal{N}$	Set of information sources (sensors, agents, Bayesians)
N	Number of information sources in system
ϵ	Generalized error symbol
$s(\cdot)$	Score function
∇	Vector Gradient
$\mathcal{A}$	Action set
a_l	lth action
a^{ir}	sensor i's rth preference action
$U(\cdot)$	Utility function
$L(\cdot)$	Loss function
$\beta(\cdot)$	Expected Utility
$\mathcal{B}(\cdot)$	Total expected utility
p_l	Outcome probability distribution function for action a_l
$\rho_l, \{p\}_l$	Set of probabilistic outcomes corresponding to action a_l
$\{\rho\}_l$	Set of probabilistic outcomes for *all* sensors in $\mathcal{N}$ corresponding to a_l
$\succeq$	Preferred to or equal in preference to
$\Rightarrow$	Implies
$h(\cdot)$	Entropy
$i(\cdot)$	Mutual information
J	Fisher information
$\mathcal{I}(\cdot)$	General information metric
I	Entropy based posterior information metric

General Symbols(cont)

Symbol	Interpretation
i	Entropy based likelihood information metric
$\tilde{I}$	Entropy based partial information metric
$\mathcal{T}$	Target or feature set
r	Range to feature
σ_r^2	Range variance
θ	Bearing to feature
σ_θ^2	Bearing variance
δ	Differential path length
Δ	Differential time of flight
β_{RCD}	Angular width of Region of Constant Depth
r_{RCD}	Range of Region of Constant Depth
β_{min}	Angular criteria
ϵ_r	Range error criteria
ζ	Sensor control parameters
d	Separation (base-line) of Polaroid devices on Tracking Sonar
σ_δ^2	Observation noise variance in differential estimator filter
σ_u^2	State transition noise in differential estimator filter
$\mathcal{F}\{\cdot\}$	Fourier transform

Abbreviations

AEP	Asymptotic Equipartition Property
CCD	Charged Coupled Device
DKF	Decentralized Kalman filter
DSN	Distributed Sensor Networks
EKF	Extended Kalman Filter
FOV	Field of View
JPDAF	Joint Probabilistic Data Association Filter
KF	Kalman Filter
MAP	Maximum a posteriori, estimate

Nomenclature

ML	Maximum likelihood, estimate
MMSE	Minimum mean square error, estimate
NNSF	Nearest Neighbour Standard Filter
OxNav	Oxford Navigator research vehicle
PDF	Probability distribution function
PDAF	Probabilistic Data Association Filter
RCD	Region of constant depth
R	Receiver
SPMS	Single Platform Multi-Sensor
SVD	Singular Value Decomposition
T	Transmitter
TOF	Time-of-flight

1 Introduction and Overview

Knowledge is of two kinds, we know a subject ourselves, or we know where we can find information upon it.
- Dr Samuel Johnson. 1775

In mammals, sensory perception provides a way of satisfying the need for knowledge concerning the external environment. Similarly, while an autonomous robot system may have some *a priori* knowledge about its environment, in order to gain knowledge about the external environment and the robot's relation to that environment or to update and refine such knowledge, sensing techniques are employed. Therefore, sensory perception can be defined broadly as the process of obtaining and maintaining an internal description of the external world. In sensing, as in the information age we find ourselves in, the problem is often not one of a shortage of information but rather one of making sense of diverse and vast amounts of it, or, of recognising that which is relevant and useful. An intuitive approach is to seek and extract only information that is necessary and relevant to the task at hand. Such information is then combined and the appropriate inferences drawn from it. This simple approach forms the underlying theme which permeates the work presented in this book.

1.1 Sensing and Multi-Sensor Systems

For our purposes here, we define a robot as an autonomous system which, as a pre-requisite to carrying out its functions, must gain knowledge about its environment in order to make inferences about itself in relation to that environment.

Sensors are used extensively in robotics to tackle the problem of perception by providing information about the external world. Sen-

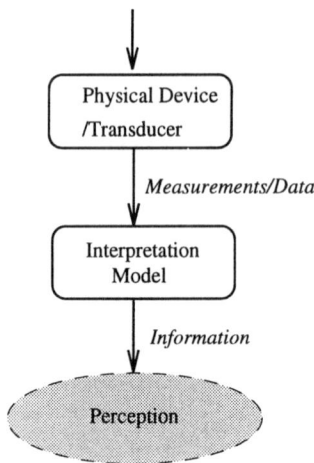

Figure 1.1: Sensing. From a sensor device measurement to perceptual information.

sors exploit physical phenomena to measure quantities. The measured quantities are expected to provide information about the *state of nature*. In this case, the state of nature refers to whatever quantities, parameters or variables are of perceptual interest to the robot system. A particular sensor device is considered appropriate for a sensing task when a relationship or mapping exists between the measured quantity and the state of nature. The exactitude with which this relationship is known depends on how well understood the measurement is, in as far as it relates to the state. In this regard, physical descriptions of sensors are invariably useful. However, such descriptions or *physical models* are unavoidably only approximations owing to our lack of complete understanding of the principles governing the transducer operation and consequently the resulting measurement. This is often exacerbated by incomplete knowledge and understanding of the environment and its interaction with the sensor. In addition, sensor measurements inherently incorporate varying degrees of uncertainty and are, occasionally, spurious and incorrect. This, coupled with the practical reality of occasional sensor failure greatly compromises reliability and reduces confidence in sensor measurements. Also, the spatial and physical limitations of sensor devices often mean that only *partial* information[1] can be provided

---

[1]Abidi [3] likens sensors to narrow passband filters capable of passing only partial information about the state of interest.

1.1 Sensing and Multi-Sensor Systems

by a single sensor.

As a result of these shortcomings, a single sensor has limited capabilities for resolving ambiguities and providing consistent descriptions of the sensed environment. And so, despite advances in sensor technologies and the myriad computational methods and algorithms aimed at extracting as much information as possible from a given sensor, the irrefutable fact remains; no single sensor is capable of obtaining *all* the required information *reliably*, at all times, in different and sometimes dynamic environments. It is thus clear that the sensing functionality needed in complex robotic systems far exceeds the repertoire of any single sensor. Motivated by biological organisms, which in essence are multi-sensory perception systems, intelligent robotic systems make use of a multiplicity of sensors in order to extract as much information as possible about a sensed environment.

Multi-sensor systems find application in areas ranging from process plant monitoring and control, surveillance, to military command and control systems. Multi-sensor systems aim to overcome the shortcomings of single sensors by employing:

- **Redundancy.** Redundancy is the use of two or more sensors to measure the same or overlapping quantities or spaces. It is well known that redundancy reduces uncertainty. This can be appreciated from the fact that for two sensors, the signal relating to the measured quantity is correlated, whereas the uncertainty associated with each sensor tends to be uncorrelated. Also, redundancy is desirable if sensor failure is anticipated so that system performance is degraded gracefully.

- **Diversity and Complementarity.** Physical sensor diversity is based on the use of different sensor technologies together. Spatial diversity offers differing viewpoints of the sensed environment simply by having sensors in different locations. Such diversity is extremely useful in efforts to reduce uncertainty and is invaluable in resolving ambiguities. Complementarity results if the sensor suite is made up of sensors each of which observes a subset of the environment state space, such that the union of these subsets makes up the whole environment state space which is of perceptual interest to the robot system.

Examples of multi-sensor systems are that described by Mitchie and Aggarwal [154] which obtains complementary information from visual, thermal and range sensors and that described by Flynn [82] which combines sonar and infra-red sensors. The literature is replete with examples; the work by Allen and Bajcsy [5] using vision and touch, Elfes [74] using sonar and stereo range data, Nandhakumar and Aggarwal [162] using thermal and visual images, Terzopolous [200] using spatially diverse visual information. The general trends towards the use of multi-sensor systems are reported in the surveys by Giralt [87], Luo [141] and Waltz and Llinas [203].

The theory and application of multi-sensor systems is determined and defined by the approaches adopted in order to address the following fundamental issues:

1. How can the diverse and sometimes conflicting information provided by sensors in a multi-sensor system, be combined in a consistent and coherent manner and the requisite states or perceptual information inferred?

2. How can such systems be optimally configured, utilised and co-ordinated in order to provide, in the best possible manner, the required information in often dynamic environments?

These are the fundamental issues which *Sensor Data Fusion* addresses.

1.2 Data Fusion and Sensor Management

Clearly, what sensor data fusion encompasses depends on how it is defined. In a literal sense, *data* refers to the actual measurements taken or information obtained by the sensors (sources) and *fusion* is the process of combining this data or information in such a way that the result provides more information than the sum of the individual parts. Figure 1.2 illustrates a data fusion problem in which multiple uncertain measurements z_i are taken from a feature by each sensor i in order to obtain a single estimate of the state of the feature.

Definitions of sensor data fusion, as employed in research literature, vary in scope albeit with same theme. This is illustrated by the following representative definitions:

> "Data fusion is the process by which data from a multitude of sensors is used to yield an optimal estimate of a specified state vector pertaining to the observed system." Richardson and Marsh [183].

1.2 Data Fusion and Sensor Management

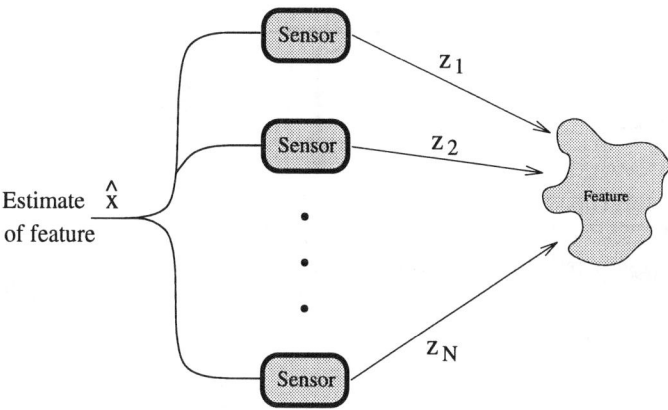

Figure 1.2: Sensor data fusion.

"... the problem of *sensor fusion* is the problem of combining multiple measurements from sensors into a single measurement of the sensed object or attribute, called the *parameter*." McKendall and Mintz [151].

"Data fusion deals with the synergistic combination of information made available by various knowledge sources such as sensors, in order to provide a better understanding of a given scene." Abidi and Gonzales [2].

"*Multisensor fusion*, ..., refers to any stage in the integration process where there is an actual combination (or fusion) of different sources of sensory information into one representational format." Luo [140].

Of these definitions, that of Abidi and Gonzalez is the most comprehensive as it incorporates the notion of *Sensor Synergy*. Sensor synergy can be described as the organisation, coordination and management of sensors and the combination of the information they provide such that their overall operation is complementary and non-conflicting given the sensing needs of the system. This synergistic operation has also been termed *Sensor Management* [203][172], Sensor coordination [68][96] and Sensor planning or control [95][40]. We use the term Sensor Management to mean the process which seeks to manage or coordinate the use of sensing resources in a manner that improves the process of data fusion and ultimately that of perception, synergistically. The alternate term

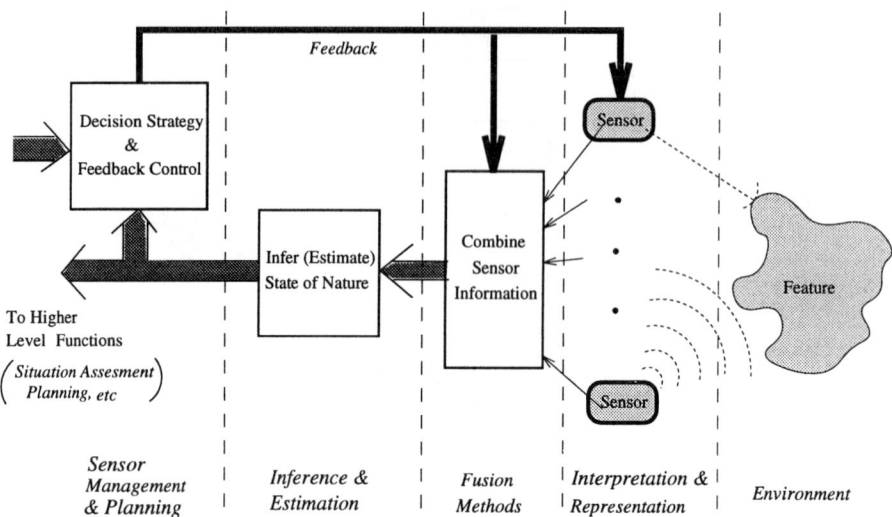

Figure 1.3: A taxonomy of issues in Multi-Sensor Data Fusion.

Managed Data Fusion, perhaps better captures both the concepts of data fusion and sensor management.

1.2.1 A Taxonomy of Issues

Irrespective of the specifics of a given application, the issues which are pervasive in sensor data fusion can, in abstraction, be summarized as follows (see Figure 1.3):

- **Interpretation and Representation.** Firstly, sensor measurements need to be interpreted and understood. This is based on a detailed description of the physical nature of the measurement and a geometrical description of the sensed environment. In addition, the way in which the measurement relates to the state of nature must be well understood. Secondly, as a consequence of physical sensor diversity, the measurements obtained are varied. This diversity of measurements covers the whole gamut, for example from range information to light intensities. Thus, it is expedient to have a common method of representing or describing the information contained in the measurements. The representation should encapsulate sensor uncertainty and performance as related to the observation of the state of nature.

1.2 Data Fusion and Sensor Management

- **Fusion, Inference and Estimation.** Required are methods for combining diverse information in a manner that is consistent, coherent and avoids distortion or biases caused by malfunctioning sensors or outlier (rogue) measurements. Having combined all the available relevant information, we need to infer the state of nature in a manner that is optimal by some criteria, given the inherent uncertainty in sensor measurements. Methodologies can be presented based on structural paradigms such as architectures, logical structures or as estimation based algorithms. Such methods are facilitated by encoding sensor information in a common representation.

- **Sensor Management.** When presented with several sensing options or configurations, the option making the best use of sensor resources to achieve system goals must be chosen. Implicit is a requirement for an *a priori* understanding of the perceptional goals of the system. These goals provide a basis for the criteria used to evaluate the efficacy of alternative strategies and configurations.

While these issues are fundamental in any data fusion problem, a major consideration which determines the form of methods employed, is the multi-sensor *architecture*.

1.2.2 Background

The main issues in sensor data fusion have mostly been addressed separately, sometimes based on well-founded theories and sometimes in an ad hoc manner and in the context of specific systems and architectures. We now discuss some of this work with respect to each of the issues summarized above. The methods applied are, in general, either quantitative or qualitative. However, we emphasize quantitative methods because quantitative analysis provides a common language and embodies a corpus of well-understood techniques while also providing a means of performance evaluation. Nevertheless, mention will be made, where pertinent, of qualitative methods. This background is intended as a preamble to the approach that this monograph presents and thus reflects a bias towards that approach.

Sensor Models and Representation

Sensor models are, in the first instance, aimed at the interpretation of measurements and are therefore based on developing an understanding of the sensor measurements and the sensed environment. Such an approach to sensor modelling is exemplified in the models presented by Kuc and Siegel [127] and Leonard [134] for sonar and Hillis [107] for a touch sensor. Models of this form, while invaluable in providing good physical and geometrical descriptions of sensors, do not quantitatively describe the uncertainty of the sensor measurements. In addition, measurements are not described in a common representation which can be made use of in a general multi-sensor system.

Probability theory is the most widely used method for describing and representing uncertainty in a way that abstracts from a sensor's physical and operational details. Durrant-Whyte [70, 69] develops uncertainty geometry and addresses the representation problem by developing probabilistic models and describes such models as "... enabling different sensors to communicate to each other in a common dimensionless language". Such quantitative methods have been used by Faugeras and Ayache [75], among others, to evaluate and model uncertainty in vision sensing. More recently, probabilistic models for sensors have been developed by Barshan [18] for several inertial navigation sensors, based on techniques described by Jenkins and Watts [118]. Probabilistic descriptions are extremely useful in augmenting physical models as shown in Manyika and Durrant-Whyte [145, 147] thus providing a way of objectively evaluating sensors and the information they provide using a common language (representation). Other approaches which provide a common representation include descriptions based on the detection capabilities of sensors. These descriptions are used to classify sensors into hard-decision sensors and soft-decision sensors as described by Waltz and Llinas [203]. However, such approaches are restricted to sensors that are used for detection and classification as explored by Nahin [160] for parametric data and by Buede and Waltz [38].

Qualitative methods (see Cohn [50]), have also been used to describe sensors as exemplified by Flynn [82] for sonar and infra-red and the logical sensor descriptions of Henderson [105]. Quantitative descriptions are useful when it is intended to address data fusion using qualitative reasoning methods or when used to prime quantitative methods.

A distinction between quantitative and qualitative methods can be made in terms of their respective descriptive capabilities. Quantitative

1.2 Data Fusion and Sensor Management

descriptions can be said to be *behavioural* in that they describe a sensor's measurements in terms of the statistics of the data and the geometry of the measured environment. Hence, a shortcoming of quantitative methods is that it is often difficult to encode the physical descriptions of the devices in order to describe variation with physical sensing parameters and environments. On the other hand, quantitative methods are *structural* in that they provide descriptions based on the physical properties of the sensing devices together with of the sensed environment. The work by Bobrow [28] discussed qualitative methods for describing physical systems. It can be appreciated that such structural descriptions, in addition to their use in priming quantitative data fusion, can be extremely useful in the selection and management of sensors, and indeed the work by Flynn [82] is an early example of this. The sensor model approach that we adopt, therefore, encompasses both physical descriptions of the sensing process as well as geometrical and statistical descriptions of the data.

Sensor Fusion and Inference

Much work has been done in developing methods of combining information from different sensors. The basic approach has been to pool the information using what are essentially "weighted averaging" techniques of varying degrees of complexity. Stone [199] proposes such a pool based on a probabilistic representation of information. There are several variations on the theme and Berger [23] discusses these while Zidek [210] provides a survey of these methods. An initial discussion of sensor probabilistic fusion can be found in Durrant-Whyte [67]. Geometric methods, which are a slight variation on the theme yielding similar results, have been proposed for sensor fusion by Nakamura [161] and Lee [131]. Non-probabilistic methods such as Dempster Shafer evidential reasoning [209] have also been used for fusion as described by Garvey *et al* [85]. Bogler [29] describes applications of non-probabilistic methods to multi-sensor target identification. Inferring the state of nature given a probabilistic representation is, in general, a well understood problem in classical statistical estimation (see Gelb [86]). Representative are methods such as Bayesian estimation [23][171], Least Squares estimation, Kalman Filtering [16][196] etc. Application of the Kalman filter and its derivatives has been widely reported; typical is the work by Willner [60], Hashemipour *et al* [104], Willsky *et al* [208], Castanon and Teneketzis [41] and by Durrant-Whyte *et al* [177, 91] where it is used for both fusion and inference. McKendall and Mintz have addressed fusion by employing statistical decision theory based on a class of robust minimax decision rules [151].

Qualitative methods for inferring the state using measurements from disparate sources have also been developed [85] and include methods which make use of Neural networks [128] and Expert systems [5]. In [101], Harris discusses artificial intelligence approaches in military systems.

Sensor Management

Sensor planning or management is invaluable as a means of actively reducing uncertainty and resolving ambiguities. To date, sensor management has mostly been addressed using descriptive techniques as made evident in the survey by Waltz [203] of several implemented systems. Much of the work done in sensor management has been in the area of tracking radar systems, for example the work by McKenzie and Mullens [152] and Weinberg [207]. In these systems, the approach has been to develop models of sensor behaviour and performance and then manage sensors on that basis. This approach is facilitated by the centralized or hierarchical nature of these systems. Ikeuchi and Kanade [115] describe a system which automatically generates sensor control programs based on feature detection performance. More formal attempts to manage sensors are presented in the work described in [194][99] and more recently [123]. A large proportion of sensor allocation schemes are based on determining cost functions and performance trade-offs *a priori*, as described by Hovanessian [111] and later by Balchen and Dessen [13]. Nash [164] uses cost-payoff assignment matrices to allocate sensors to targets, while Rothman and Bier[24] make use of a boolean matrix which defines sensor-target assignments based on sensor availability and capability. Expert system approaches have also been used such as that simulated by Cowan [53] which demonstrates the benefits to data fusion of managing sensors and the simulation for a distributed surveillance system by Leon and Heller [133]. Donald and Jennings [65] develop methodology based on equivalence classes for interpreting sensors and their data and subsequently directing sensing actions so as to obtain information necessary to complete a manipulation task.

A demonstrably rational way of making decisions such as those required for sensor management, is through the use of *normative* or decision-theoretic techniques. Normative techniques are desirable be-

1.2 Data Fusion and Sensor Management

cause they are based on an axiomatic framework and have a performance that can be analysed objectively. Making decisions in this way is hard [23] because of the required rigour and the associated computational difficulties in finding the optimal action and as a result this is a research area in its own right [206]. Tsitsiklis and Athans [202] highlight these difficulties for a distributed detection problem. Efforts have been made to use normative methods in the pre-selection of sensors as exemplified in the work by Hager [95] and Fleskes [81] and in the active pursuit of uncertainty reduction by Hager [94]. A particularly relevant example is the work by Blackman [25] that makes use of a utility approach for real-time sensor selection where the utility is related to the predicted covariance matrix in a Kalman filter. Most functional implementations of managed data fusion systems are in combat aircraft where achieving sensor synergy is of the essence given the stringent timing, operational and performance constraints. Such systems are described in [203] and [172] and make use of a mixture of the techniques already mentioned.

Architectures

Architectures have traditionally been centralized, utilizing a central processor responsible for implementing data fusion. The need to relieve computational burdens at the central processor leads to hierarchical systems [167][48] which have the advantage of allowing for several levels of abstraction and indeed processing. Whilst ideal for coordination and control, hierarchical architectures are vulnerable to processor failure, computational bottlenecks and inflexibility. The need to overcome these shortcomings and the recent trends towards *autonomous* systems have led to the development of distributed architectures and Distributed Sensor Networks (DSNs). Such architectures are described by Chong [49] and, more generally, Iyengar [116] provides a recent survey of DSNs. A refinement of these are the *Decentralized* architectures described by Hashemipour *et al* [104], Tsitsiklis and Athans [202] and Durrant-Whyte [72, 71] and specified in Manyika and Durrant-Whyte [144]. In summary, decentralized systems offer:

- **Modularity.** This is a result of the fact that local sensing and global data fusion takes place at the sensor node itself. This is facilitated by each sensor node being fully autonomous with its own sensing, processing and communication facilities.

- **Scalability and Flexibility.** Because all the functionality is localized in the sensor, scaling the system is simply a matter of adding or removing sensors.

- **Survivability.** Due to the absence of a central processor, the system can withstand the loss of nodes and performance is gracefully degraded.

Summary

The above survey illustrates the eclectic nature of methods that have been developed for data fusion and sensor management. This has resulted in implemented systems using a mixture of methods[2]. There are hardly any reported systems making use of a single framework to address *all* the issues of representation, fusion and management, in a consistent manner amenable to a variety of architectures. The work of Hager [95] and that of Blackman [25] and most recently [26], comes close to this ideal although the methods are not directly applicable to decentralized systems. The same can be said of geometric methods which can be the basis of both fusion and parametric sensor planning. This monograph attempts to fill this particular gap by presenting a single approach which is applicable, consistently, to a variety of architectures and applications.

1.3 An Information-Theoretic Approach

It has already been alluded to that, in robotics, multi-sensor systems are used as the primary means of perception. Our approach is predicated on perception being concerned, primarily, with acquiring information related to the state of nature. This makes the acquisition and refinement of information, the main goal of multi-sensor systems. Ergo, sensing (for purposes of perception) reduces to the simple process whereby knowledge or information concerning a given state is updated based on a sensor observation containing *relevant* information about the state. Implicit in this is an assumption that the observation can be modelled

---

[2]As examples: The system described by Kuczewski [128] uses a Neural network to manage Kalman filter based multi-target tracking in a hierarchical system. Popoli [172] describes a fuzzy decision tree approach utilizing heuristic criteria to manage multi-target tracking and emission control.

1.3 An Information-Theoretic Approach

in terms of its information value, which requires detailed *a priori* knowledge of the source and nature of the observation. This basic paradigm can be refined to take account of uncertainty, in the knowledge representation and in the observation itself, and also to take account of the availability of observations from several sources. By adopting a probabilistic approach to represent uncertainty and information, we develop a *probabilistic information update* relation within a Bayesian framework capturing the essence of sensing in multi-sensor systems. Using the probabilistic information update, we can address the issues of representation and fusion in general and for decentralized systems in particular.

Of particular importance in multi-sensor systems is sensor management. Addressing sensor management in decentralized systems presents some added difficulties. These are due to the autonomous nature of the sensor nodes in this system and the absence of a central processor (coordinator). These problems are as follows; (i) how to guarantee consistency and consensus amongst decision-makers, (ii) the nature of the criteria for optimality and the question of group or individual optimality, and not least (iii) the maintenance of coherence and rationality in the decisions made. These are the classic problems encountered in group or decentralized decision theory, where it is generally accepted that these questions can only be addressed properly using normative methods. Moreover, a normative approach to decentralized management is a reasonable and rational way to proceed because the resulting decision structure lends itself to quantitative and objective analysis. In addition, normative methods are applicable to a variety of architectures. Several hurdles are encountered in the development of a normative approach to sensor management:

1. The information upon which decisions will be based must be formally described and modelled to ensure objectivity and rationality in the decision making. And, in a decision-theoretic sense, this suggests a probabilistic description of the sensor information.

2. A basis for preference and optimality is required such that decisions and their outcomes may be evaluated accordingly.

3. In general, the computations required for normative methods can be considerably complex [172].

While generally prescribed solutions for decentralized decision theory have thus far been elusive [23][206], for sensing systems certain assumptions can be made which simplify the problem. Not least of which is that the information from each sensor can be modelled objectively. And since the purpose of the system is to gain information about an environment, this information can itself be the basis for evaluating actions and decisions.

In general, the concept of information can itself be a rather nebulous one, often defying definition. Fortunately, in the realm of probabilities and stochastic systems, it is possible to develop quantitative definitions of information. And for this we look to Information Theory and the measurement of uncertainty in probability distributions. Owing to our use of an information paradigm, elements of information theory and decision-theoretic methods we term our approach *Information-Theoretic*.

In summary, using the probabilistic information-theoretic framework, we develop and present the following:

1. **A Probabilistic Data Fusion Model.** This model provides the probabilistic elements necessary for the theoretical development of data fusion and sensor management methods within a single consistent framework.

2. **Data Fusion Algorithms and Architectures.** Building on the probabilistic data fusion model, algorithms for inference in multi-sensor systems are derived, together with the architectures for implementing them.

3. **Information Metrics.** These are metrics used to evaluate and compare the information available in a multi-sensor system in general, and in decentralized systems in particular.

4. **Normative Sensor Management Methods.** These methods are based on a decentralized decision-theoretic approach, making use of information based utility functions.

1.4 Application and Practical Demonstration

The survey in Section 1.2.2 gives a flavour of the enormous variety of multi-sensor applications. Indeed, the approach which we put forward in this book is applicable to a wide variety of multi-sensing problems. In our own work we have encountered applications ranging from surveillance systems to process plants. In order to illustrate the effectiveness

1.4 Application and Practical Demonstration

of our approach we have chosen to concentrate on a particular application and demonstrations associated with this application. The choice of this application has in part been influenced by our current research interests and also by the illustrative nature of the results. We apply the theoretical work to address problems in mobile robot navigation, specifically, the problems of multi-sensor vehicle localization and feature extraction for map-building.

It has been suggested by Leonard [134] and others, that vehicle localization on the basis of a map or, conversely, map-building on the basis of location information, can be greatly improved by using directed sensing strategies such as *focus of attention*. Focus of attention aims at reducing the overall quantity of data processed while increasing its informational value by providing only correctly associated measurements [3]. If applied to localization, where a sensor on a vehicle focusses attention on a known feature in the environment, the high band-width of correctly associated measurements from such a sensor can be used *directly* to determine the location of the vehicle. This is in contrast to current methods of obtaining large amounts of measurements, correlating these with map information each localization cycle and subsequently using only correctly correlated measurements to estimate location. The uncorrelated measurements are not utilised and can be discarded. When applied to map-building, a sensor with focus of attention capability can provide information relating to an un-mapped feature which can then be used to incorporate the feature into the map. Thus, a vehicle equipped with such sensors is capable of both localizing and map-building simultaneously, given an initial knowledge of some environment feature(s).

Implementation of navigation in this way is, however, assumes the availability of sensors with the ability to focus attention on given geometrical features such as the ones naturally occurring in indoor environments. Towards this end and also in order to demonstrate practical techniques for sensor modelling, we outline a model for a *differential sonar* sensor which is similar to the *monopulse* principle used in radar. Exploitation of the differential principle results in a Tracking Sonar capable of focussing attention on a given feature while producing correctly-associated measurements at rates of up to 30Hz. Such a sensor provides information which can be used for localization and feature extraction.

---

[3]Focus of attention can be found in natural vision systems such as in saccade eye motion. Vision research has attempted to mimic this as described by Andersson [8] and Sharkey *et al* [192].

Thus, a navigating mobile robot can be equipped with several such sensors in a decentralized configuration. Such a system provides a good application platform for the methods that we develop for data fusion and sensor management. Indeed, through this application we are able to address all the various issues raised in Figure 1.3. More specifically, we employ the decentralized data fusion algorithms to fuse the information from the tracking sonar sensors and estimate the location of the vehicle. We also implement a feature classification algorithm which complements the tracking of un-mapped features for map-building.

The need for sensor management in such an implementation is self-evident, given that each sensor can track any one of several features in its field of view. Decisions, therefore, need to be made, firstly, about which features need to be tracked and, secondly, in what sensor configuration, at a given moment. This is necessary in order to obtain the most useful information for the given task that the sensors are performing, be it localization or feature extraction. In addition, it may become necessary as the vehicle moves around to assign and re-assign features (targets) to sensors and to cue sensors and hand-off features as they move in and out of view.

The work described is implemented on purpose-built vehicles which have been designed by the OxNav group at Oxford. OxNav vehicles making use of the above elements have been demonstrated in several practical indoor environments[4].

In summary, our application and implementation are intended to demonstrate the following:

1. **A Sensor Model.** The detailed modelling of a sensor in terms of its operation and the probabilistic information that it provides, as required by our normative approach. Here we also experimentally justify the observation noise assumption for our specific sensor.

2. **Decentralized Directed Navigation Algorithms.** A scheme for decentralized localization and feature classification based on algorithms presented. The term *directed* is due to the fact that the sensors are directed to seek and obtain only that information which is necessary.

3. **Sensor Management.** Demonstrate how sensor management functions such as sensor-feature assignments, sensor cueing and hand-off can be achieved on a navigating robot decentrally.

---
[4]For a rather amusing third party account of this work see *New Scientist* [1].

1.5 A Guided Tour of the Book

We commence in **Chapter 2** by presenting the various elements of a probabilistic model for data fusion and sensor management. We present a Bayesian model of the observation process for a sensor and the subsequent inference of the state. We then consider the problem of combining probabilistic information from several sources. We present Bayesian decision-theory and the ancillary utility theory. Finally, we introduce information-theoretic considerations.

In **Chapter 3**, we are concerned with developing architectures and algorithms for data fusion. We present information filters for estimating a continuous state, culminating in the decentralized information filter. We complement this algorithm with an equivalent one for the classification of discrete states.

Chapter 4 presents a normative approach to data fusion management. We construct utility functions based on information as expected utility. We present entropy and Fisher information metrics for the information provided by data fusion. Using these metrics, we develop methods for making the decisions necessary for decentralized management.

Having thus far emphasized the crucial nature of sensor models, we address, in **Chapter 5**, precisely that issue for sonar. We explain sonar measurements based on a Region of Constant Depth (RCD) model. We present a differential sonar arrangement which leads to the development of a Tracking Sonar with a focus of attention capability. More importantly, we develop a probabilistic model of the information provided by the sensor.

In **Chapter 6**, we apply the work presented in the preceding chapters to problems in mobile robot navigation. We use the algorithms of Chapter 3 and the sensor model of Chapter 5 to address vehicle localization and feature classification. This implementation also highlights the need for sensor management and coordination.

Chapter 7 presents some demonstrations of sensor management for a navigating robot. Results are presented for several sensor-feature assignment problems during localization and feature classification. Sensor hand-off and cueing are also demonstrated.

We conclude in **Chapter 8** by summarizing the significance of a consistent approach to managed data fusion. More importantly, we highlight the limitations and potential developments based on this work, with a view towards developing general paradigms for data fusion. Practical concerns with regards to fully functional systems are also discussed.

Summary: *A Road Map*

Figure 1.4 summarizes the work presented in this monograph in the form of a *Road Map* through the chapters. The map is intended to show the progression of the concepts presented [5].

---

[5] As Sir Peter Medawar points out [153], the actual process of scientific thought and research exhibits no such ordered progression (such as in the Road Map) and that, in this regard, published scientific works are a kind of fraud or misrepresentation. This is because the actual order in which events and ideas are developed is eradicated from the published work due to the desire and indeed requirement to present ideas in a clear and consistent progression. This is certainly the case here.

1.5 A Guided Tour of the Book

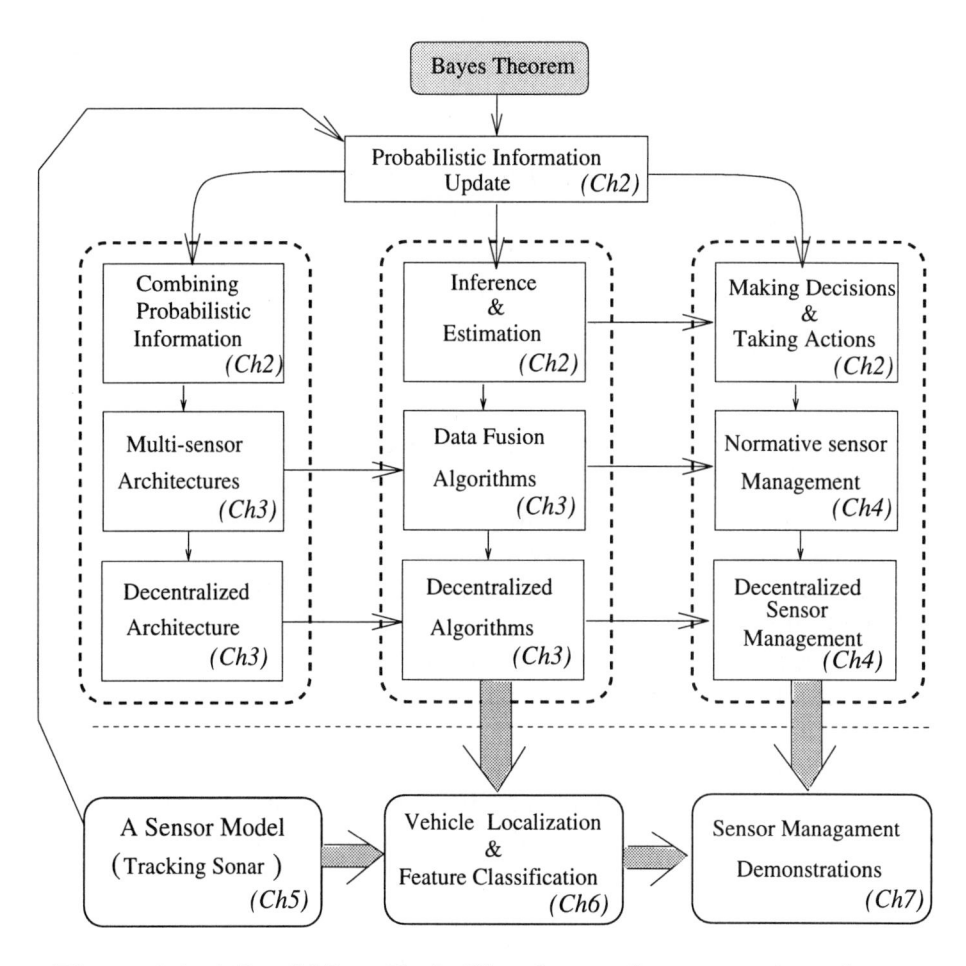

Figure 1.4: A Road Map. Each *Ch* refers to the appropriate chapter.

2 A Probabilistic Model for Managed Data Fusion

The importance of probability can only be derived from the judgement that it is rational to be guided by it in action; and a practical dependence on it can only be justified by a judgement that in action we ought to act to take some account of it.

- John Maynard Keynes (A Treatise on Probability)

In a data fusion problem, we commence by determining a state of nature which we are interested in. Such a state may be a description of the spatial location of an object, its identity in terms of attributes, a complex dynamic state or simply a single numeric quantity. The objective is to infer the true state based on, often incomplete and sometimes conflicting, information obtained from a variety of sources and also to coordinate (manage) and optimize the way in which such information[1] is acquired. By providing measurement data, a sensor can be viewed abstractly as an *information source*. In a multi-sensor system several such information sources are available, thus making it possible to implement different strategies for obtaining and combining information. This necessitates the development of methods for evaluating and making decisions regarding alternate strategies.

Probabilistic descriptions are indispensable when representing and dealing quantitatively with uncertainty. Indeed, probability theory is rich enough and complete in itself to the extent that, based upon it, issues in managed data fusion can be addressed in a rational and consistent manner. Hence, this chapter presents a probabilistic information-theoretic model for managed data fusion. In essence, what is presented are elements making up a framework from which methodologies can be

---

[1]Until we define information in Section 2.4, we shall use the term to mean data, observed or otherwise, which is related to the state of nature.

developed for addressing data fusion and sensor management in multi-sensor systems. The presentation brings together elements of probability theory, Bayesian methods, estimation theory, decision theory and information theory. These elements are the basis for addressing; (i) uncertainty and the sensor observation process, (ii) the fusion of diverse sensor information, (iii) the estimation and inference of the underlying state of nature and (iv) the making of decisions necessary for sensor management.

2.1 Observations, Inference and Estimation

We introduce some notation and formalism in order to facilitate the ensuing discussion. The state of nature is described by an n-dimensional vector x taken from some state space $\mathcal{X}$, that is, $\mathbf{x} \in \mathcal{X}$. The state vector is denoted

$$\mathbf{x} = [x_1, x_2, \ldots, x_n]^T, \tag{2.1}$$

and may be discrete or continuous. When the state is continuous, the space $\mathcal{X} \subseteq \Re^n$, is the n-dimensional Euclidean space. The state vector x may also be random or deterministic and in what follows we make no distinction unless otherwise stated. We take m measurements which give some indication as to the state x. These measurements make up the m-dimensional observation vector z taken from some observation space $\mathcal{Z}$, that is, $\mathbf{z} \in \mathcal{Z}$. In our context, where we make physical measurements represented by real numbers, the observation space is such that $\mathcal{Z} \subseteq \Re^m$. Each measurement may itself be random or deterministic. Usually $m \geq n$, however, when $n > m$, the measurements then only provide partial information and in some cases it may not be possible to extract *all* the required information about x from these measurements alone.

The observation vector is related to the state by

$$\mathbf{z} \triangleq \mathcal{M}(\mathbf{x}, \mathbf{v}), \tag{2.2}$$

where v is an unknown observation noise vector of the same dimensionality as z and is normally described by a random variable. $\mathcal{M}$ is a generalized *observation model* relating the state space $\mathcal{X}$ to the observation space $\mathcal{Z}$. $\mathcal{M}$ is never known precisely and invariably a practical model of $\mathcal{M}$ always incorporates some approximations. The closeness with which such a model approaches the "true" relation between $\mathcal{Z}$ and $\mathcal{X}$ requires a good understanding of the phenomenological nature of the observations, a physical model of the measurement device itself and knowledge of the sensed environment.

The noise associated with the sensor observations and the approximations in the observation model result in considerable uncertainty in information concerning the state. It is useful to gain an insight into the nature of this uncertainty.

2.1.1 A Geometrical Representation of Uncertainty

Sensor fusion is motivated by a need to gain information about x, and an intrinsic part of this is the need to reduce uncertainty about the state x. Insight into this can be obtained by considering a geometric interpretation of the uncertainty in x. From the relation in Equation 2.2, we can observe that the state that we infer is a function of the observations and their inherent uncertainty, i.e.

$$\mathbf{x} = \mathbf{f}(\mathbf{z} + \mathbf{v}), \tag{2.3}$$

in which we have assumed the measurement noise v to be additive. By also assuming v to be small, we can approximate the above as

$$\mathbf{x} \approx \mathbf{f}(\mathbf{z}) + \nabla_{\mathbf{z}}\mathbf{f}\mathbf{v}, \tag{2.4}$$

where $\nabla_{\mathbf{z}}\mathbf{f}$ is the Jacobian matrix of the function f with respect to z. From this, it can be shown easily that the uncertainty (strictly the covariance) in x is represented by

$$\nabla_{\mathbf{z}}\mathbf{f} E\{\mathbf{v}\mathbf{v}^T\} \nabla_{\mathbf{z}}\mathbf{f}^T, \tag{2.5}$$

where $E\{\mathbf{v}\mathbf{v}^T\}$ is the covariance matrix of the noise v. For uncorrelated noise v, the matrix $\nabla_{\mathbf{z}}\mathbf{f} E\{\mathbf{v}\mathbf{v}^T\} \nabla_{\mathbf{z}}\mathbf{f}^T$ is symmetric and hence its singular value decomposition (SVD) is given by

$$\nabla_{\mathbf{z}}\mathbf{f} E\{\mathbf{v}\mathbf{v}^T\} \nabla_{\mathbf{z}}\mathbf{f}^T = (\mathbf{U}\mathbf{D}\mathbf{U}^T), \tag{2.6}$$

where U is an orthogonal matrix of dimension $(n \times n)$ made up of vectors $\mathbf{e}_j$, that is

$$\begin{aligned} \mathbf{U} = (\mathbf{e}_1, \ldots, \mathbf{e}_n), \quad \text{such that } \mathbf{e}_k^T \mathbf{e}_j &= 1, \text{ for } j = k \\ &= 0, \text{ for } j \neq k, \end{aligned} \tag{2.7}$$

and

$$\mathbf{D} = diag(d_1, \ldots, d_n), \quad \text{where } d_1 \geq \ldots \geq d_n \geq 0. \tag{2.8}$$

2.1 Observations, Inference and Estimation

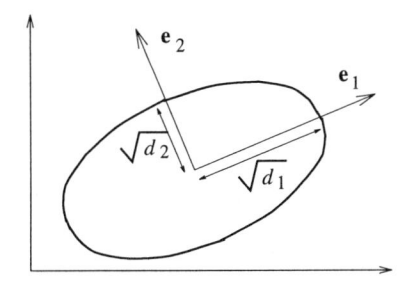

Figure 2.1: Uncertainty ellipsoid for a 2-dimensional state.

The scalar variance in each direction corresponding to each of the components of x is given by the corresponding component of **D**. When all the directions are considered for a given state x, the geometrical result is an ellipsoid with principal axes in the directions $e_j, \forall j$, with $2\sqrt{d_j}$ as the corresponding magnitudes. The volume of the ellipsoid thus represents the uncertainty in x. Figure 2.1 shows the ellipsoid for a 2 dimensional state. When $n > m$, it turns out that the uncertainty ellipsoid has infinitely long principal axes in the directions corresponding to elements in x for which there is no information contained in z.

Therefore, in a geometric sense, the aim is to reduce the volume of the uncertainty ellipsoid. Towards this end, we develop methods for data fusion and inference and estimation. In fact it can be shown [161] that the classical methods for inference and estimation do reduce the uncertainty ellipsoid. The above representation of uncertainty is the basis for various methods widely used for validation, representation of uncertainty in state estimates and indeed in parametric sensor planning. In order to develop the methods which reduce the above uncertainty, we turn to the methods which originate from the 18th century mathematician Reverend T. Bayes.

2.1.2 A Bayesian Paradigm

The probabilistic information contained in z about x is described by the probability distribution function (PDF) $p(\mathbf{z} \mid \mathbf{x})$, known as the *likelihood function*. Such information is considered *objective* because it is based only on observations. The likelihood function contains all the relevant information from the observation z required in order to make inferences about the true state x. This is termed the *Likelihood Principle*. It can be shown using the factorisation theorem[2] as derived by Lehman [132], that the likelihood function is proportional to a function $g(T(\mathbf{z}) \mid \mathbf{x})$, i.e.

$$p(\mathbf{z} \mid \mathbf{x}) \propto g(T(\mathbf{z}) \mid \mathbf{x}), \qquad (2.9)$$

where $T(\mathbf{z})$ is a *sufficient statistic* of $\mathbf{x}$. The statistic $T(\mathbf{z})$ is said to be sufficient for the unknown state $\mathbf{x}$ if the conditional distribution $p(\mathbf{z} \mid T(\mathbf{z}))$ is independent of $\mathbf{x}$. The sufficient statistic summarizes all the information contained in the observation. Therefore, from statistical theory, such a sufficient statistic can be used to make inferences and decisions related to the state $\mathbf{x}$. Using this notion of sufficiency, the raw observed data can be discarded and a minimal sufficient statistic used to make inferences.

The likelihood function does not, however, tell the whole story if, before measurement, information about the state $\mathbf{x}$ is made available exogenously. Such *a priori* information about the state can be encapsulated as the prior distribution function $p(\mathbf{x})$ and is regarded as *subjective* because it is not based on any observed data. How such prior information and the likelihood information interact to provide *a posteriori* information, is solved by Bayes Theorem which gives the posterior conditional distribution of $\mathbf{x}$ given $\mathbf{z}$ as

$$p(\mathbf{x} \mid \mathbf{z}) = \frac{p(\mathbf{z} \mid \mathbf{x})\, p(\mathbf{x})}{\int p(\mathbf{z} \mid \mathbf{x}) p(\mathbf{x})\, d\mathbf{x}} = \frac{p(\mathbf{z} \mid \mathbf{x}) p(\mathbf{x})}{p(\mathbf{z})}, \qquad (2.10)$$

where $p(\mathbf{z})$ is the marginal distribution. An equivalent formulation, which often proves quite useful, arises from a consideration of sufficient statistics as follows; if $T(\mathbf{z})$ is a sufficient statistic for $\mathbf{x}$ with distribution $g(T(\mathbf{z}) \mid \mathbf{x})$, then, from the factorization theorem, we can write the posterior as

$$p(\mathbf{x} \mid \mathbf{z}) = p(\mathbf{x} \mid T(\mathbf{z})) = \frac{g(T(\mathbf{z}) \mid \mathbf{x})\, p(\mathbf{x})}{\int g(T(\mathbf{z}) \mid \mathbf{x}) p(\mathbf{x})\, d\mathbf{x}}. \qquad (2.11)$$

Often the distribution $g(T(\mathbf{z}) \mid \mathbf{x})$ is easier to handle and model than the actual likelihood $p(\mathbf{z} \mid \mathbf{x})$ and so in such a case, Equation 2.11 would be used instead of Equation 2.10. If no subjective prior information is available, then a distribution $p(\mathbf{x})$ is required in Equation 2.10 such that the resulting posterior contains no more information than that

---

[2]Sometimes known as the Fisher-Neyman factorisation.

2.1 Observations, Inference and Estimation

contained in the likelihood. Such a prior distribution is called a *non-informative prior*, the discussion of which is deferred until Section 2.4.2.

In an attempt to reduce uncertainty and resolve ambiguity, several measurements may be taken over time before constructing the posterior. The measurements are usually taken in discrete time and we shall denote each time-step as k. We define the set of all observations up to time k as follows

$$\mathbf{Z}^k \triangleq \{\mathbf{z}(1), \mathbf{z}(2), \cdots, \mathbf{z}(k)\}. \tag{2.12}$$

The posterior distribution of $\mathbf{x}$ given the set of observations $\mathbf{Z}^k$ is now computed as

$$p(\mathbf{x} \mid \mathbf{Z}^k) = \frac{p(\mathbf{Z}^k \mid \mathbf{x})p(\mathbf{x})}{p(\mathbf{Z}^k)}, \tag{2.13}$$

or can be computed recursively after each observation $\mathbf{z}(k)$ as follows

$$p(\mathbf{x} \mid \mathbf{Z}^k) = \frac{p(\mathbf{z}(k) \mid \mathbf{x})\, p(\mathbf{x} \mid \mathbf{Z}^{k-1})}{p(\mathbf{z}(k) \mid \mathbf{Z}^{k-1})}. \tag{2.14}$$

With this recursive definition, we are not required to store all the observations, but need only consider the current observation $\mathbf{z}(k)$ at each kth step. This is the form of Bayes Theorem which is most commonly used in practice. In this form the value of the Bayesian paradigm, that it provides a direct and easily applicable means of combining observed information with prior beliefs, is made explicit. Given this, the pervasiveness of Bayes Theorem in data fusion problems is unsurprising.

2.1.3 Classical Inference and Estimation

Inference and estimation of the underlying state of nature is at the heart of data fusion problems. Given the inherent uncertainty in the system, the problem is one of obtaining the best possible estimate of the state. In the notation introduced above, inference is concerned with obtaining an estimate $\hat{\mathbf{x}}$ of the state of nature from the posterior distribution $p(\mathbf{x} \mid \mathbf{Z}^k)$ together with some measure of estimation accuracy. Consequently, an estimator is considered *optimal* if it minimizes the probability of error based on some criterion.

An intuitive approach to estimation is to find the value of $\mathbf{x}$ in $\mathcal{X}$ which is most likely based on the information available. Such information is in the form of probability distributions and hence classical methods of estimation maximize some formulation of this probabilistic information. Classical inference techniques include the *Maximum Likelihood* (ML) estimate obtained by maximizing the likelihood function, i.e. information obtained from observations only

$$\hat{\mathbf{x}}_{ML} = \arg\max p(\mathbf{Z}^k \mid \mathbf{x}). \qquad (2.15)$$

For a general distribution, the ML estimate corresponds to the mode of the distribution. The *Maximum a posteriori* (MAP) estimate is obtained by maximizing the posterior distribution i.e. information obtained from observations as well as other prior information

$$\hat{\mathbf{x}}_{MAP} = \arg\max p(\mathbf{x} \mid \mathbf{Z}^k). \qquad (2.16)$$

Since prior information is quite often subjective, objectivity in an estimate (or inferred state) is said to be maintained by considering only observed information, i.e. the likelihood function. Therefore, an ML estimate is said to be objective. Correspondingly, a MAP estimate obtained using a non-informative prior is also considered to be objective. This is because such a MAP estimate is identical to an ML estimate.

A fundamental concept in statistical estimation is that of *expectation*. We define $E^{p(\mathbf{x})}$ or simply E, the expected value of some function $f(\mathbf{x})$ with respect to a distribution $p(\mathbf{x})$ as

$$\begin{aligned} E\{f(\mathbf{x})\} &= \int f(\mathbf{x})p(\mathbf{x})\,d\mathbf{x}, \quad \text{if continuous,} \\ &= \sum_{\mathbf{x}\in\mathcal{X}} f(\mathbf{x})p(\mathbf{x}), \quad \text{if discrete.} \end{aligned} \qquad (2.17)$$

The *moments* of a distribution can be obtained using the expectation operator. The posterior mean (1st moment) and covariance (2nd moment) are given by

$$\bar{\mathbf{x}} \triangleq E^{p(\mathbf{x}|\mathbf{Z}^k)}\{\mathbf{x}\}, \qquad (2.18)$$

and

$$\mathbf{P}_{\mathbf{x}|\mathbf{Z}^k} \triangleq E^{p(\mathbf{x}|\mathbf{Z}^k)}\left\{(\mathbf{x}-\bar{\mathbf{x}})(\mathbf{x}-\bar{\mathbf{x}})^T\right\}, \qquad (2.19)$$

respectively. It can be shown that over all possible estimates $\hat{\mathbf{x}}$, the posterior mean $\bar{\mathbf{x}}$ minimizes the estimate covariance as follows;

$$\begin{aligned} \mathbf{P}_{\hat{\mathbf{x}}} &= E\left\{(\mathbf{x}-\bar{\mathbf{x}}+\bar{\mathbf{x}}-\hat{\mathbf{x}})(\mathbf{x}-\bar{\mathbf{x}}+\bar{\mathbf{x}}-\hat{\mathbf{x}})^T\right\} \\ &= E\left\{(\mathbf{x}-\bar{\mathbf{x}})(\mathbf{x}-\bar{\mathbf{x}})^T\right\} + E\left\{2(\mathbf{x}-\bar{\mathbf{x}})(\bar{\mathbf{x}}-\hat{\mathbf{x}})^T\right\} \\ &\quad + E\left\{(\bar{\mathbf{x}}-\hat{\mathbf{x}})(\bar{\mathbf{x}}-\hat{\mathbf{x}})^T\right\} \\ &= \mathbf{P}_{\mathbf{x}|\mathbf{Z}^k} + (\bar{\mathbf{x}}-\hat{\mathbf{x}})(\bar{\mathbf{x}}-\hat{\mathbf{x}})^T, \end{aligned}$$

2.1 Observations, Inference and Estimation

from which setting $\nabla_{\hat{\mathbf{x}}} \mathbf{P}_{\hat{\mathbf{x}}} = 0$ yields

$$\nabla_{\hat{\mathbf{x}}} \mathbf{P}_{\hat{\mathbf{x}}} = -2\bar{\mathbf{x}} + 2\hat{\mathbf{x}} = 0, \quad \text{therefore,} \quad \bar{\mathbf{x}} = \hat{\mathbf{x}}. \qquad (2.20)$$

Showing that $\nabla_{\hat{\mathbf{x}}} \nabla_{\hat{\mathbf{x}}}^T \mathbf{P}_{\hat{\mathbf{x}}} > 0$, demonstrates that the covariance $\mathbf{P}_{\hat{\mathbf{x}}}$ is minimized when the estimate equals the posterior mean. Estimators which minimize the posterior covariance are termed *minimum variance* estimators.

A particularly popular minimum variance estimator is the Least Squares Estimate which minimizes the sum of the square errors, that is, minimizes the Euclidean distance between the true state $\mathbf{x}$ and the estimate $\hat{\mathbf{x}}$, given the set of observations $\mathbf{Z}^k$. The equivalent estimator for random variables is called the *Minimum Mean Square Error* (MMSE) estimate and is written

$$\hat{\mathbf{x}}_{MMSE} = \arg\min_{\hat{\mathbf{x}}} E^{p(\mathbf{x}|\mathbf{Z}^k)} \left\{ (\hat{\mathbf{x}} - \mathbf{x})(\hat{\mathbf{x}} - \mathbf{x})^T \right\}. \qquad (2.21)$$

Differentiating with respect to $\hat{\mathbf{x}}$ and equating to zero gives the minimizing estimate i.e.

$$\nabla_{\hat{\mathbf{x}}} \int (\mathbf{x} - \hat{\mathbf{x}})^T (\mathbf{x} - \hat{\mathbf{x}}) p(\mathbf{x} \mid \mathbf{Z}^k) \, d\mathbf{x} = -2 \int (\mathbf{x} - \hat{\mathbf{x}}) p(\mathbf{x} \mid \mathbf{Z}^k) d\mathbf{x} = 0$$

$$\text{and so,} \quad \hat{\mathbf{x}} = \int \mathbf{x} p(\mathbf{x} \mid \mathbf{Z}^k) d\mathbf{x} = E\{\mathbf{x} \mid \mathbf{Z}^k\}. \qquad (2.22)$$

Thus, an MMSE estimate is given by the conditional mean. That $\hat{\mathbf{x}}$ is an MMSE estimate can also be shown geometrically by a decomposition of the error between $\mathbf{x}$ and an arbitrary estimator of $\mathbf{x}$, $\mathbf{g}(\mathbf{z})$ into 2 orthogonal components, followed by a consideration of the mean-squared error between $\mathbf{x}$ and the arbitrary estimator. The mean-squared error is given by

$$E\left\{(\mathbf{x} - \mathbf{g}(\mathbf{z}))^T (\mathbf{x} - \mathbf{g}(\mathbf{z}))\right\} \geq E\left\{(\mathbf{x} - \hat{\mathbf{x}})(\mathbf{x} - \hat{\mathbf{x}})^T\right\}, \qquad (2.23)$$

with equality iff $\mathbf{g}(\mathbf{z}) = \hat{\mathbf{x}}$. It follows from the previous discussion that the MMSE estimate is also a minimum variance estimate. When the mean of the conditional density $p(\mathbf{x} \mid \mathbf{Z}^k)$ coincides with the mode, the MAP estimate is equivalent to the MMSE estimate.

In [161], it is shown that these methods and their derivatives such as the Kalman filter, do in fact reduce the uncertainty ellipsoid associated with the state x. Using these concepts, estimation algorithms can be developed for various applications.

The techniques presented thus far are, in general, well understood in terms of classical statistical theory. However, when there is a multiplicity of information sources in a variety of configurations and topologies, a number of more complex issues arise. In the next and subsequent sections, we consider the combination of information from several diverse sources.

2.2 Combining Probabilistic Information

In the previous section, the use of multiple observations from a single source was presented as a way of reducing uncertainty. Another approach aimed at reducing uncertainty and obtaining more complete knowledge of the state of nature, is the *fusion* of information originating from a number of spatially and physically diverse sources. In tackling the problem, the following questions should be considered:

- How *relevant* to the problem at hand is information from each source?

- How *reliable* is the information from each source?

Addressing the first question entails validating or checking the information from i for its relevance as regards inference and decisions related to x. The second question can be addressed by attaching a measure of value such as a weight to the information provided by each source depending on *a priori* reliability information if this is known. Making the assumption (similar to Assumption 1, page 37), for now, that these issues have been taken account of, we can consider the problem of combining information.

We define $\mathcal{N}$, the set of all the available information sources (sensors)

$$\mathcal{N} = \{i\}, \quad \text{for } i = 1, 2, \ldots, N. \tag{2.24}$$

The most intuitive method for sensor fusion, assuming sensors which make direct measurements $\mathbf{x}'_j$ of the state x, is to simply obtain a weighted average i.e.

$$\hat{\mathbf{x}} = \sum_j w_j \mathbf{x}'_j, \tag{2.25}$$

2.2 Combining Probabilistic Information

where $\sum_j w_j = 1$, and the summations over $\mathcal{N}$. There are shortcomings with this rather simplistic approach. Firstly, sensors make measurements z which may not be directly related to x (see Equation 2.2). However, this can be taken into account by incorporating an observation model in Equation 2.25. Secondly, and most importantly, the simplistic method does not take account of the uncertainty inherent in sensor observations and so the resulting estimate $\hat{x}$ does not incorporate any notion of optimality in a statistical sense.

A way to proceed from this quandary is to consider fusion in a probabilistic sense. This builds on the probabilistic description of the sensing and inference process that we have described thus far. The initial difficulty lies in the method for combining probabilistic information. To formalize the development of probabilistic information fusion, we start by defining the set of *all* observations made by the set of sensors $\mathcal{N}$ up to time-step k as

$$\{\mathbf{Z}^k\} = \cup_i \mathbf{Z}_i^k, \quad \forall i \in \mathcal{N}, \qquad (2.26)$$
$$\text{where } \mathbf{Z}_i^k = \{\mathbf{z}_i(1), \mathbf{z}_i(2), \cdots, \mathbf{z}_i(k)\},$$

and $\mathbf{Z}_i^k$ is information source i's set of observations up to the time-step k. What is now required is to compute the global posterior distribution $p(\mathbf{x} \mid \{\mathbf{Z}^k\})$, given the information contributed by each source. In what follows, we will assume that each information source communicates either a local posterior PDF $p(\mathbf{x} \mid \mathbf{Z}_i^k)$ or a likelihood $p(\mathbf{z}_i(k) \mid \mathbf{x})$ (assuming the recursive form of Equation 2.14).

2.2.1 Pooling Information

We consider in turn the two approaches generally proposed in the literature, and discuss some of the criticisms associated with them from a sensor fusion perspective. Our discussion culminates in what we have termed the Independent Likelihood Pool.

Linear Opinion Pool

The linear opinion pool as described by Stone [199] is a deceptively simple approach to aggregating probability distributions. The posteriors from each information source are combined linearly i.e.

$$p(\mathbf{x} \mid \{\mathbf{Z}^k\}) = \sum_j w_j \, p(\mathbf{x} \mid \mathbf{Z}_j^k), \qquad (2.27)$$

where w_j is a weight such that, $0 \leq w_j \leq 1$ and $\sum_j w_j = 1$. This is illustrated in Figure 2.2. The weight w_i reflects the significance attached

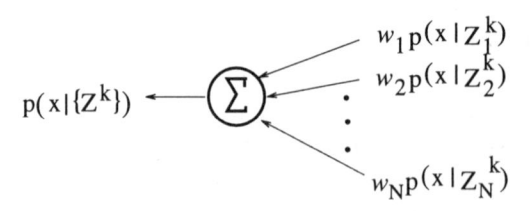

Figure 2.2: Linear Opinion Pool.

to information source i. The weights can also be used to model the reliability or trustworthiness of an information source and to "weight out" faulty sensors. However, the formulation of Equation 2.27 requires that weight w_i be known before information from i is evaluated, and this presents some difficulty. Though a general methodology for obtaining the weights w_i has not been forthcoming, problem specific methods have been developed. Worthy of mention is the entropy-based weighting technique by Basir and Shen [20] which establishes the level of dependence between information sources and then computes appropriate weights before information aggregation.

Applying Bayes theorem to Equation 2.27 and assuming each information source has its own subjective prior distribution gives

$$p(\mathbf{x} \mid \{\mathbf{Z}^k\}) = w_1 \frac{p(\mathbf{Z}_1^k \mid \mathbf{x})p(\mathbf{x}_1)}{p(\mathbf{Z}_1^k)} + w_2 \frac{p(\mathbf{Z}_2^k \mid \mathbf{x})p(\mathbf{x}_2)}{p(\mathbf{Z}_2^k)} + \cdots + w_N \frac{p(\mathbf{Z}_N^k \mid \mathbf{x})p(\mathbf{x}_N)}{p(\mathbf{Z}_N^k)}.$$

Such addition of probabilities (assuming equal weights) results in an inability to reinforce opinion. This is because the Linear Opinion Pool does not take appropriate cognizance of the possible availability of independent information locally at each node i. An illustrative example is provided by considering a discrete state $\mathbf{x} = \{x_1, x_2\}$.

Example. If $(N-1)$ information sources report posteriors of $(0.1, 0.9)$ and one dissenting source j reports $(0.7, 0.3)$, for a large N it seems sensible for j to be ignored or reinforced towards $(0.1, 0.9)$. The following are the results for a 5 sensor system

$$\frac{1}{5}\sum_5 (0.1,\ 0.9) = (0.1,\ 0.9), \tag{2.28}$$

2.2 Combining Probabilistic Information

and when one sensor is dissenting

$$\frac{1}{5}\left((0.7,\ 0.3) + \sum_4 (0.1,\ 0.9)\right) = (0.22,\ 0.78). \quad (2.29)$$

The Linear Opinion Pool lends undue credence to j's opinion and so gives an erroneous result. The need to redress this leads to the second approach.

Independent Opinion Pool

The Independent Opinion Pool is defined by the product

$$p(\mathbf{x} \mid \{\mathbf{Z}^k\}) = \alpha \prod_j p(\mathbf{x} \mid \mathbf{Z}_j^k), \quad (2.30)$$

where α is a normalizing constant. This is illustrated in Figure 2.3.

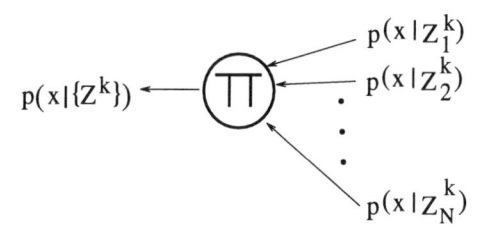

Figure 2.3: Independent Opinion Pool.

The implicit assumption in the Independent Opinion Pool is that the information obtained conditioned on the observation set is independent. In general, this is a difficult condition to satisfy particularly when the information sources are human! However, in the realm of measurement and experimentation based on physical laws and principles, the conditional independence of the observations can often be justified experimentally. This is usually done in the process of modelling the likelihood function $p(\mathbf{z}(k) \mid \mathbf{x})$ by showing that the residual uncertainty in each observation arises from uncorrelated noise terms (more on this in Chapter 5). The independent pool allows for the reinforcement of opinions which, when each source is assumed to have subjective prior information, is appropriate. However, the Independent Opinion Pool is extreme in its reinforcement of opinion when prior information at each node is common i.e. obtained from the same source. This can be seen by expanding and rewriting the global posterior distribution as

$$p(\mathbf{x} \mid \{\mathbf{Z}^k\}) = \alpha \left[\frac{p(\mathbf{Z}_1^k \mid \mathbf{x}) p(\mathbf{x}_1)}{p(\mathbf{Z}_1^k)} \times \frac{p(\mathbf{Z}_2^k \mid \mathbf{x}) p(\mathbf{x}_2)}{p(\mathbf{Z}_2^k)} \times \cdots \times \frac{p(\mathbf{Z}_N^k \mid \mathbf{x}) p(\mathbf{x}_N)}{p(\mathbf{Z}_N^k)} \right]. \tag{2.31}$$

If the prior information is obtained from the same source, then

$$p(\mathbf{x}_1) = p(\mathbf{x}_2) = \cdots = p(\mathbf{x}_N), \tag{2.32}$$

and this results in unwarranted reinforcement of the posterior through the product of the priors in Equation 2.31 i.e. $\prod_j p(\mathbf{x}_j)$. As such, the Independent Opinion Pool is only appropriate when the priors are obtained independently on the basis of subjective prior information at each information source.

2.2.2 Independent Likelihood Pool

When each information source has common prior information i.e. information obtained from the same origin, the opinion pool which more accurately describes the situation is developed as follows: Bayes theorem for the global posterior, that is, the distribution of $\mathbf{x}$ conditioned on *all* the observations up to time k, is given by

$$\begin{aligned} p(\mathbf{x} \mid \{\mathbf{Z}^k\}) &= \frac{p(\{\mathbf{Z}^k\} \mid \mathbf{x}) \, p(\mathbf{x})}{p(\{\mathbf{Z}^k\})} \\ &= \frac{p(\mathbf{Z}_1^k, \mathbf{Z}_2^k, \ldots, \mathbf{Z}_N^k \mid \mathbf{x}) \, p(\mathbf{x})}{p(\mathbf{Z}_1^k, \mathbf{Z}_2^k, \ldots, \mathbf{Z}_N^k)}. \end{aligned} \tag{2.33}$$

The distribution $p(\mathbf{Z}_1^k, \mathbf{Z}_2^k, \ldots, \mathbf{Z}_N^k \mid \mathbf{x})$, is difficult to compute in practice if there are dependencies, which do not depend on $\mathbf{x}$, among the elements of each information source's observation set and those of other information sets. However, for sensor systems, it is reasonable to assume that the likelihoods from each information source i, that is, $p(\mathbf{Z}_i^k \mid \mathbf{x})$ are independent. This is because the only parameter that the observations have in common is the state and even so, we must show in practice that

$$p(\mathbf{Z}_1^k, \mathbf{Z}_2^k, \cdots, \mathbf{Z}_N^k \mid \mathbf{x}) = p(\mathbf{Z}_1^k \mid \mathbf{x}) p(\mathbf{Z}_2^k \mid \mathbf{x}) \cdots p(\mathbf{Z}_N^k \mid \mathbf{x}). \tag{2.34}$$

2.2 Combining Probabilistic Information

If this is the case, we can write Equation 2.33 as

$$\begin{aligned}p(\mathbf{x} \mid \{\mathbf{Z}^k\}) &= \frac{p(\mathbf{Z}_1^k \mid \mathbf{x})\, p(\mathbf{Z}_2^k \mid \mathbf{x}) \cdots p(\mathbf{Z}_N^k \mid \mathbf{x})\, p(\mathbf{x})}{p(\mathbf{Z}_1^k, \mathbf{Z}_2^k, \ldots, \mathbf{Z}_N^k)} \\ &= \frac{p(\mathbf{x}) \prod_j p(\mathbf{Z}_j^k \mid \mathbf{x})}{p(\mathbf{Z}_1^k, \mathbf{Z}_2^k, \ldots, \mathbf{Z}_N^k)},\end{aligned} \qquad (2.35)$$

which can be written recursively as

$$p(\mathbf{x} \mid \{\mathbf{Z}^k\}) = \alpha\, p(\mathbf{x} \mid \{\mathbf{Z}^{k-1}\}) \left[\prod_j \underbrace{p(\mathbf{z}_j(k) \mid \mathbf{x})}_{\text{likelihood}} \right], \qquad (2.36)$$

where α is a normalizing constant independent of x. The Indepen-

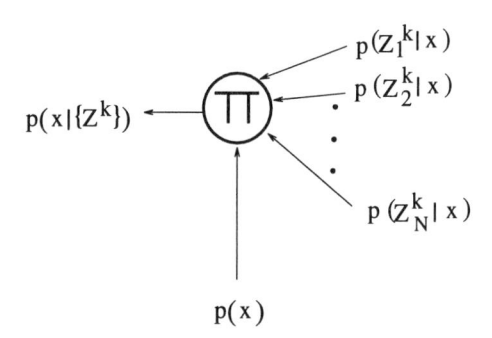

Figure 2.4: Independent Likelihood Pool.

dent Likelihood Pool is illustrated in Figure 2.4. From a communication standpoint, the Independent Likelihood Pool is consistent with the Bayesian update of Equation 2.14 as follows: We can write an expanded observation set $\{\mathbf{Z}^k\}$ according to Equation 2.27 which encompasses all the observations from all the sources and then simply write Equation 2.14 in terms of this expanded observation set as if the observations were from a single source.

2.2.3 Remarks

Of the methods described above, the Independent Opinion Pool and the Independent Likelihood Pool more accurately model the situation in multi-sensor systems where the conditional distributions of the observations can be shown to be independent. The choice between these pools depends on the origin of the prior information. For most situations in sensing, the Independent Likelihood Pool is the most appropriate method because of the prior information in these systems tends to have the same origin. The Linear Opinion Pool is useful if there are dependencies between information sources and it is known *a priori* how to assign individual weights to the probabilistic information from each source.

Of particular importance in this book is the architectural paradigm which results when the combination of probabilistic information is distributed. Each information source performs the tasks of combining information from all the other information sources and computing a global posterior, with the *proviso* that all the information sources in the system are able to communicate with each other directly. This amounts to replicating the independent opinion or likelihood pools (Equations 2.27-2.36) at every information source with some simplifications resulting. In this way, the posterior obtained is identical at each information source [3].

Inference and estimation in multi-information source systems is based on the same methods as described in Section 2.1.3. Instead of using the posterior $p(\mathbf{x} \mid \mathbf{Z}^k)$, we make use of the global posterior $p(\mathbf{x} \mid \{\mathbf{Z}^k\})$. The MAP and MMSE estimators discussed previously can be obtained from the global posterior. However, for the ML we make use of

$$p(\{\mathbf{Z}^k\} \mid \mathbf{x}) = \prod_j p(\mathbf{Z}^k_j \mid \mathbf{x}), \qquad (2.37)$$

based on the independence assumption made earlier.

A note on Data Association

Data association is the problem of correctly associating data or measurements with the correct sub-vector or subspaces of the state. This occurs when each observation provides information concerning only a sub-vector or subspace of the state. In this book we do not directly address data association in detail, however, the problem can still be addressed using a Bayesian approach [198]. As a simple illustration, consider a sensor system in which the sub-vectors of the state are $\mathbf{x}_1, \ldots, \mathbf{x}_n$.

---

[3] Our choice of combining information in such a distributed system using the independent opinion and likelihood pools differs from the approach by Chair and Varshney [44] which is based on distributing a form of the Linear Opinion Pool which they proposed in [43].

2.3 Decisions and Actions

Measurements $z_1, \ldots, z_n$ are made and it is known that each z_i corresponds to a sub-vector x_{j_i}. Therefore, the set $\{j_1, \ldots, j_n\}$ is a permutation of the set $\{1, \ldots, n\}$. Assuming we are able to obtain the PDFs $p(z_i \mid x_j)$, $\forall i, j$, we can find the most probable association for each observation z_i. We start by computing the posterior which results from each possible association i.e.

$$p(\mathbf{x} \mid \mathbf{z}) = \frac{p(z_i \mid x_j) p(x_j)}{\int p(z_i \mid x_j) p(x_j) dx}, \quad \forall i, j = 1, \ldots, n. \quad (2.38)$$

Therefore, for each z_i we find the most probable association, in a MAP sense, by finding the index $\hat{j}$ which maximizes the posterior, i.e.

$$\arg\max_j p(x_j \mid z_i). \quad (2.39)$$

The estimate of each x_j then follows using methods that we have already discussed.

2.3 Decisions and Actions

Managing sensor systems involves making decisions and taking actions under conditions of uncertainty. The decisions under consideration in sensor management are those concerned with evaluating the next best sensing operation, sensor-feature assignments and sensor resource allocation etc. Modelling the sensing process probabilistically makes the problem amenable to techniques developed in Decision Theory [77][206][95]. Our starting point in making such decisions is the *a posteriori* distribution, $p(\mathbf{x} \mid \mathbf{Z}^k)$ in the case of a single information source and, in the case of multiple information sources, $p(\mathbf{x} \mid \{\mathbf{Z}^k\})$ and its various local components. The posterior PDF encompasses all that is known about the state $\mathbf{x}$ and, therefore, decisions concerning $\mathbf{x}$ can justifiably be based on it.

2.3.1 Decisions and the Single Bayesian

We define an action a which is an element of a set of possible actions $\mathcal{A}$. We also define a *Utility function* $U(\mathbf{x}, a)$, defined for all $(\mathbf{x}, a) \in (\mathcal{X} \times \mathcal{A})$, which gives a measure of the utility of taking action a when the true state is a particular $\mathbf{x}$. Conversely, we can define a *Loss function* $L(\mathbf{x}, a)$ which gives a measure of the loss incurred in taking action a when the state is **x**. Hence, it can be written that $U(\mathbf{x}, a) = -L(\mathbf{x}, a)$. The expected utility (or loss) β of taking an action a can be computed as follows

$$\beta(p(\mathbf{x} \mid \mathbf{Z}^k), a) \triangleq E^{p(\mathbf{x}|\mathbf{Z}^k)}\{U(\mathbf{x}, a)\}. \tag{2.40}$$

The Bayes action $\hat{a}$ is the strategy which maximizes the expected utility i.e.

$$\hat{a} = \arg\max_{a} \beta(p(\mathbf{x} \mid \mathbf{Z}^k), a). \tag{2.41}$$

From the Likelihood Principle, it can be shown that this is equivalent to maximizing

$$\int_{\mathbf{x}} U(\mathbf{x}, a)\, p(\mathbf{Z}^k \mid \mathbf{x})\, p(\mathbf{x})\, d\mathbf{x}. \tag{2.42}$$

Analytically, this is a rational framework for making decisions and taking actions, the caveat being an ability to determine appropriate utility functions.

When the action space is the same as the state space the decision problem becomes similar to the inference problem already discussed. For instance, a loss function such as the squared-error loss can be defined such that the minimization of the loss form of Equation 2.40 results in an alternate statement of the MMSE estimate which we have already encountered. The minimization can be written as

$$\min_{\hat{\mathbf{x}} \in \mathbf{x}} \int L(\hat{\mathbf{x}}, \mathbf{x}) p(\mathbf{x} \mid \mathbf{Z}^k)\, d\mathbf{x},$$

where the loss L is defined as the squared error loss

$$\underbrace{L(\hat{\mathbf{x}}, \mathbf{x}) = (\hat{\mathbf{x}} - \mathbf{x})^2}_{\text{mmse}}, \quad \text{where } \hat{\mathbf{x}} = \int \mathbf{x}\, p(\mathbf{x} \mid \mathbf{Z}^k)\, d\mathbf{x}. \tag{2.43}$$

Having already introduced inference, we shall not consider this any further and instead concern ourselves only with action spaces which are not the same as the state space.

2.3.2 Decisions involving Multiple Information Sources

When considering decision-making with multiple sources of information, it is important to delineate and define the problem precisely in terms of the structural level of the decision-making, the nature and level of interaction during the decision-making and the presence or absence of a dominant decision maker. We make the following assumption:

2.3 Decisions and Actions

Assumption 1 : *Individual information sources (Bayesians) are coherent, cooperate fully and are able to communicate probabilistic information as required with negligible delay.*

The problem of making decisions involving multiple information sources has been formulated under a variety of assumptions in the literature [27][139]. We shall consider the following representative cases:

Case 1: Consider a system consisting of N information sources and a single overall decision maker. Such a system is illustrated in Figure 2.5. The task of the decision maker is, in the first instance, to combine probabilistic information from all the sources and then make decisions based on the global posterior. Given that the global posterior is $p(\mathbf{x} \mid \{\mathbf{Z}^k\})$, the Bayes group action is given by

$$\begin{aligned}
\hat{a} &= \arg\max_{a} \beta(p(\mathbf{x} \mid \{\mathbf{Z}^k\}), a) \\
&= \arg\max_{a} E^{p(\mathbf{x}\mid\{\mathbf{Z}^k\})} \{\mathbf{U}(\mathbf{x}, a)\},
\end{aligned} \quad (2.44)$$

where $\mathbf{U}(\mathbf{x}, a)$ is a group utility function. The solution in this case is well defined in terms of classical Bayesian analysis and is often termed the "super Bayesian" approach [206].

Case 2: Consider a system consisting of N Bayesians each able to obtain its own probabilistic information which it shares with *all* the other Bayesians before computing a posterior PDF. From the previous discussion, the posterior PDF obtained by each Bayesian i is identical and is given by $p(\mathbf{x}_i \mid \{\mathbf{Z}^k\})$. Each Bayesian is then required to compute an optimal action which is consistent with those of the other Bayesians. This system is illustrated in Figure 2.6.

This second case is the fully decentralized one and presents a considerable challenge for which a universally prescribed solution is not possible partly because of the lack of generally applicable criteria for determining rationality and optimality, coupled with the question of whether to pursue group or individual optimality. The trivial case exists where the optimal action $\hat{a}_i$ at each Bayesian i happens to be the same for all the Bayesians and thus becomes the group action. In general it is not be the same, and so a solution proceeds as follows; each Bayesian i computes an acceptable (admissible) set of actions $A_i \subseteq \mathcal{A}$ and if the set $A = \cap_j A_j$ is non-empty, then the group action is selected from the set A.

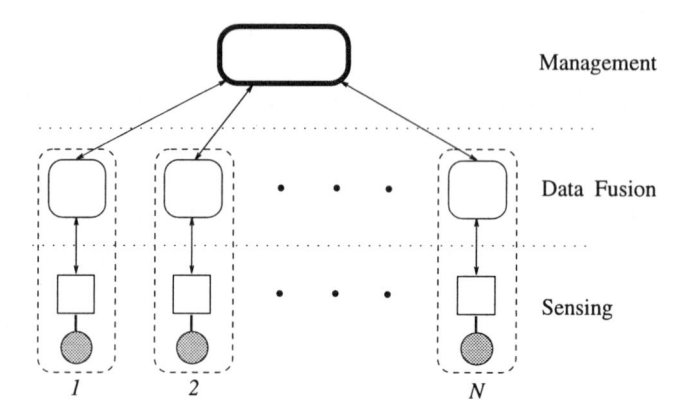

Figure 2.5: "Super" Bayesian approach to management decision-making.

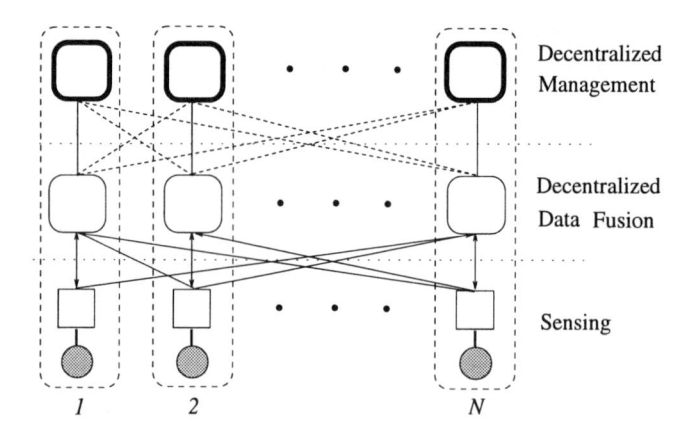

Figure 2.6: Multi-Bayesian approach to management decision-making. The connections between the decision-making and data fusion stages of different Bayesians are shown as dotted lines because they may or may not exist depending on communication requirements for decision-making.

2.3 Decisions and Actions

An acceptable class of actions can be obtained by maximizing

$$\beta(p(\mathbf{x} \mid \{\mathbf{Z}^k\}), a) = \sum_j w_j \, E^{p(\mathbf{x}|\{\mathbf{Z}^k\})} \{U_j(\mathbf{x}, a)\}, \qquad (2.45)$$

where $0 \le w_j \le 1$ and $\sum_j w_j = 1$. Equivalently, from the Likelihood Principle, this can be written as a maximization of

$$\sum_j w_j \int U_j(\mathbf{x}, a) \, p(\mathbf{Z}_j^k \mid \mathbf{x}) \, p(\mathbf{x}) \, d\mathbf{x}, \qquad (2.46)$$

using the non-recursive form of the information update Equation 2.13. That Equation 2.45 and Equation 2.46 represent acceptable actions is easy to see. What becomes extremely difficult, and has been the subject of much research, is the choice of an optimal action $\hat{a}$ from the acceptable set. If the utility functions U_i can be directly compared then maximization of Equation 2.45 or Equation 2.46 gives the optimal group action. The general validity of direct utility comparisons is questionable given Arrow's impossibility theorem [10]. However, as Savage [187] argues, there are situations where such comparisons seem plausible.

When comparisons can be justified, a general solution is obtained by maximizing Equation 2.45, the solution of which happens to be a special case of a more general solution suggested by Weerahandi and Zidek [206] stated as the maximization

$$\hat{a} = \arg\max_a \left\{ \sum_j w_j \left[E^{p_j} \{U(\mathbf{x}, a_l)\} - c(j) \right]^\gamma \right\}^{1/\gamma}, \qquad (2.47)$$

where $c(j)$ is regarded as decision maker j's security level which plays the role of "safeguarding j's interests". The weight w_j is as given in Equation 2.45 and $-\infty \le \gamma \le \infty$. The case when $\gamma = 1$, gives a solution which minimizes Equation 2.45 and $\gamma = -\infty$ and $\gamma = \infty$ give the so-called Bayesian max-min and min-max solutions respectively [23][206]. The solution arising from using $\gamma = 1$ and $\gamma = \infty$ can result in an individual decision-maker ending up with a net loss of expected utility, depending on the value of $c(j)$.

When direct comparisons cannot be justified, the celebrated Nash [163] solution which makes no direct utility comparisons, can be used. This solution is obtained by maximizing the product of individual expected utilities

$$\hat{a} = \arg\max_a \prod_j \left[E^{p_j} \{U(\mathbf{x}_j, a_l)\} - c(j) \right], \qquad (2.48)$$

in which the value of $c(j)$ can be used to maintain some level of individual optimality. The value of $c(j)$ plays no part in the actual derivation of the Nash solution and is thus arbitrary. Weerahandi and Zidek have shown that setting $\gamma = 0$ in Equation 2.47 reduces it to a form of the Nash product known as the Nash-Kalai solution [119], thereby establishing an equivalence between these methods. An interesting feature of the Nash solution is that when decision-makers differ greatly, it is better to use a randomized decision rule. The degree of consensus, or lack of it, in this case case be measured by the Mahalanobis distance. Clearly, for management decisions, the use of randomized decision rules can result in undesirable sensor behaviour.

2.3.3 Utility Theory

Utility allows the placement of value on the results of decisions thus permitting a subjective preference ordering on actions and their outcomes. The results of decisions can be defined on the set $\Re$. Given uncertainty, these results may be probabilistic and defined on the set $\mathcal{P}$. We can say that action a_l gives rise to the state x which has a probability distribution p_l, where $p_l \in \mathcal{P}$. An alternative, perhaps more precise, way of writing p_l is by writing it as $p(\mathbf{x} \mid a_l)$ which makes the conditioning on a_l explicit. Because the distribution on x may also be conditioned on other states or may be a likelihood, we find the notation p_l, which makes the conditioning on a_l implicit, less confusing. The "value" of taking action a_l is given by the expectation

$$E^{p_l}\{U(\mathbf{x}, a_l)\}, \tag{2.49}$$

where $U(\mathbf{x}, a_l)$ is a real valued function which represents (through expected value) the decision maker's preference over the elements of $\mathcal{P}$. This has meaning if it is possible to state preferences amongst the elements of $\mathcal{P}$. If the distribution p_1 over x is preferred to p_2, we expect that the utility function $U(\mathbf{x}, a_l)$ is such that

$$E^{p_2}\{U(\mathbf{x}, a_2)\} < E^{p_1}\{U(\mathbf{x}, a_1)\}. \tag{2.50}$$

Therefore, $U(\mathbf{x}, a)$ is a utility function with the same preference structure as the decision maker. This ability to state preferences rationally is the basis of the axiomatic definition of utility.

In order for $U(\mathbf{x}, a)$ to be a valid utility function, the expected utilities must have the same preference ordering as the true preferences

2.3 Decisions and Actions

concerning the probabilistic outcomes in $\mathcal{P}$. This implies the need for a preference ordering among the elements of $\mathcal{P}$. The following are the axioms on which such a preference pattern is based:

- **Axiom 1.** If p_1 and p_2 are in $\mathcal{P}$ then either $p_1 \prec p_2$, or $p_1 \approx p_2$, or $p_2 \prec p_1$.

- **Axiom 2.** If $p_1 \preceq p_2$ and $p_2 \preceq p_3$, then $p_1 \preceq p_3$.

- **Axiom 3.** If $p_1 \prec p_2$ then $\alpha p_1 + (1-\alpha)p_3 \prec \alpha p_2 + (1-\alpha)p_3$, for $0 < \alpha < 1$.

Axiom 1 implies the ability to express preference and Axiom 2 is a transitivity requirement. Axiom 3 implies that in identical situations where p_1 and p_2 occur with the same probability, the preferred p based on some preference criterion is chosen. The Axioms presented above can be strengthened by the addition of another Axiom which implies that there is 'no heaven or hell', that is, no infinitely good or infinitely bad outcome. Based on the results by Fishburn [79] this fourth axiom is not necessary to guarantee the existence of utility functions. The sufficiency of the above three axioms is the basis of our use and interpretation of utility functions. These axioms guarantee that the decision maker's preference in $\mathcal{P}$ coincides with preferences according to the expected utilities of the elements of $\mathcal{P}$. A rigorous formal proof of the existence of utility functions is given by Von Neumann and Morgenstein [165] and Chernoff and Moses [47] give a more palatable proof of the same.

An implicit assumption in the above is that the decision-maker has knowledge of the exact distribution of an element of $\mathcal{P}$. Such exact knowledge is *not* normally available in real sensor systems because, firstly, the true state is unknown and estimates are used instead, and secondly, assumptions and approximations are made in order to deal with otherwise intractable modelling problems. Fortunately, we can take comfort in the fact that approximate knowledge of the distributions in $\mathcal{P}$ has been shown to be adequate in a proof by Ferguson [77] making use of so-called *personal probabilities* as defined in [9].

Various methods have been described for constructing utilities and examples of these are found in [121][23][62]. Utility functions can be modelled to give prescribed behaviour and preference profiles such as risk aversion, risk proneness and neutrality. These different behaviours are modelled analytically based on Jensen's inequality as follows; a risk-averse decision profile is implied by convexity[4] ($\cap$) i.e.

$$U(E\{\mathbf{x}\}, a) \geq E\{U(\mathbf{x}, a)\}. \tag{2.51}$$

Similarly, risk-prone behaviour is implied by the concavity

$$U(E\{\mathbf{x}\}, a) \leq E\{U(\mathbf{x}, a)\}, \tag{2.52}$$

and neutral behaviour follows from the linear relation

$$E\{U(\mathbf{x}, a)\} = U(E\{\mathbf{x}\}, a). \tag{2.53}$$

2.4 The Nature and Role of Information

Thus far we have been using the term information rather loosely to mean any quantity that is useful in providing a better understanding of the state of nature. The concept of information is a difficult one to describe and define broadly. However, the task of defining information is somewhat simplified in the context of communication and stochastic systems because states, whose informational values are of interest to us, are described in terms of probabilities. Probability distributions can thus be said to contain information about the underlying states which they describe. Various methods exist for quantifying probabilistic complexity and for summarizing probability distributions and these can be employed to formalize the concept of information. Formalizing and quantifying probabilistic information is useful for the purpose of making comparisons and determining the degree of uncertainty in the descriptions of underlying states.

2.4.1 Measures of Probabilistic Information

Entropy Metric

Entropy, sometimes known as Shannon information, is a fundamental and generally applicable measure of information in probability distributions. Entropy is the uncertainty associated with a probability distribution and as such gives a measure of the descriptive complexity of a PDF [5]. We shall define entropy as Shannon [191] first defined it; entropy

---

[4]Hereafter, convexity shall be taken to mean $\cap$ and concavity to mean $\cup$.
[5]Catlin [42] gives a particularly lucid and intuitive derivation of entropy.

2.4 The Nature and Role of Information

is defined as the expectation of the negative of the log-likelihood of a PDF. Here log-likelihood is simply the log of a distribution and is quite different from the likelihood function. And so we write that entropy is given by

$$h(p(\mathbf{x})) \triangleq E\{-ln\ p(\mathbf{x})\}. \qquad (2.54)$$

Entropy can be defined for the posterior distribution of x given $\mathbf{Z}^k$ at time k as follows

$$\begin{aligned} h(k) \triangleq h(p(\mathbf{x} \mid \mathbf{Z}^k)) &= E\{-ln\ p(\mathbf{x} \mid \mathbf{Z}^k)\} \qquad (2.55) \\ &= -\int p(\mathbf{x} \mid \mathbf{Z}^k)\ ln\ p(\mathbf{x} \mid \mathbf{Z}^k)\ d\mathbf{x}, \quad \text{if continuous,} \\ &= -\sum p(\mathbf{x} \mid \mathbf{Z}^k)\ ln\ p(\mathbf{x} \mid \mathbf{Z}^k), \quad \text{if discrete.} \end{aligned}$$

From this definition of entropy, an often useful measure of ignorance is given by $ln(1/p(\mathbf{x}))$ [6].

The entropy relationship for Bayes Theorem can be developed as follows; negating and taking expectations of the log of Equation 2.14 and then using the definition of Entropy in Equation 2.55 yields

$$\begin{aligned} E\left\{-ln\ [p(\mathbf{x} \mid \mathbf{Z}^k)]\right\} &= E\left\{-ln\ [p(\mathbf{x} \mid \mathbf{Z}^{k-1})]\right\} - E\left\{ln\ \left[\frac{p(\mathbf{z}(k) \mid \mathbf{x})}{p(\mathbf{z}(k) \mid \mathbf{Z}^{k-1})}\right]\right\} \\ h(k) &= h(k-1) - E\left\{ln\ \left[\frac{p(\mathbf{z}(k) \mid \mathbf{x})}{p(\mathbf{z}(k) \mid \mathbf{Z}^{k-1})}\right]\right\}. \qquad (2.56) \end{aligned}$$

This relationship demonstrates that conditioning with respect to observations reduces entropy, that is, increases information.

Mutual information is defined as the information about one variable contained in another. In an observation process, we define mutual information at time k as the information about x contained in the observation $\mathbf{z}(k)$, that is

$$i(k) = \mathbf{I}(\mathbf{x}, \mathbf{z}(k)) \triangleq E\left\{ln\ \left[\frac{p(\mathbf{z}(k) \mid \mathbf{x})}{p(\mathbf{z}(k))}\right]\right\}. \qquad (2.57)$$

Using these definitions, the entropy relationship for Bayes theorem (Equation 2.56) can be written as

$$h(k) = h(k-1) - i(k), \qquad (2.58)$$

---

[6] It can also been argued that ignorance may alternatively be measured by $(1-p(\mathbf{x}))$. This leads to an alternative definition of entropy as shown recently by Pal and Pal [169] which, although giving similar entropy values as the original definition, has different boundary properties which prove useful in some applications.

which simply states that, the entropy following an observation is reduced by an amount equal to the information inherent in the observation.

Fisher Information Metric

Fisher information gives a measure of the amount of information about x inherent in the observations $\mathbf{Z}^k$. Like entropy, Fisher information is developed from the log-likelihood. The gradient of the log-likelihood is called the *score function* s, given by

$$s(\mathbf{Z}^k, \mathbf{x}) \triangleq \nabla_{\mathbf{x}} ln\ p(\mathbf{Z}^k, \mathbf{x}) = \frac{\nabla_{\mathbf{x}} p(\mathbf{Z}^k, \mathbf{x})}{p(\mathbf{Z}^k, \mathbf{x})}. \tag{2.59}$$

By considering s as a random variable, its mean is obtained by considering its first moment

$$\begin{aligned} E\{s(\mathbf{Z}^k, \mathbf{x})\} &= \int \frac{\nabla_{\mathbf{x}} p(\mathbf{Z}^k, \mathbf{x})}{p(\mathbf{Z}^k, \mathbf{x})} p(\mathbf{Z}^k, \mathbf{x})\ d\mathbf{z} \\ &= \nabla_{\mathbf{x}} \int p(\mathbf{Z}^k, \mathbf{x})\ d\mathbf{z} = 0. \end{aligned} \tag{2.60}$$

The *Fisher information matrix* is defined as the covariance of the score function

$$\mathbf{J}(k) \triangleq E\{\nabla_{\mathbf{x}} ln\ p(\mathbf{Z}^k, \mathbf{x})\ (\nabla_{\mathbf{x}} ln\ p(\mathbf{Z}^k, \mathbf{x}))^T\}. \tag{2.61}$$

This can be rewritten as the negative expectation of the Hessian of the log-likelihood

$$\mathbf{J}(k) = -E\{\nabla_{\mathbf{x}} \nabla_{\mathbf{x}}^T ln\ p(\mathbf{Z}^k, \mathbf{x})\}. \tag{2.62}$$

For non-random parameters, the Fisher information matrix is defined in terms of the likelihood i.e.

$$\mathbf{J}(k) \triangleq -E\{\nabla_{\mathbf{x}} \nabla_{\mathbf{x}}^T ln\ p(\mathbf{Z}^k\ |\ \mathbf{x})\}. \tag{2.63}$$

Fisher information is useful in estimation; the inverse of the Fisher information matrix is the *Cramer-Rao lower bound* (CRLB) [16] which bounds the mean-squared error of any unbiased estimator of x.

The Fisher information measure has several useful properties which are analogous to those of entropy:

1. If $\mathbf{J}_1$ and $\mathbf{J}_2$ are measures of the information about x contained in $\mathbf{Z}_1^k$ and $\mathbf{Z}_2^k$ respectively, then $\mathbf{J}$, the information contained in $(\mathbf{Z}_1^k, \mathbf{Z}_2^k)$, is given by

$$\mathbf{J} = \mathbf{J}_1 + \mathbf{J}_2. \tag{2.64}$$

2.4 The Nature and Role of Information

2. For observation sets $Z_1^k, \ldots, Z_n^k$, which are independent and identically distributed and for which J is the information about x contained in each Z_i^k, then the total information in $(Z_1^k, \ldots, Z_n^k)$ is given by nJ.

The above results hold for both single and vector variables [182].

It is evident from the above that both entropy and Fisher information are additive, a result that shall be used later. Entropy is particularly attractive because it can be applied to both discrete and continuous distributions. However, for continuous state estimation problems, Fisher information can be more useful since algorithms can be developed which maximise it. Appendix A discusses the relationship between Fisher information and entropy.

Information measures can be used to address questions raised in Section 2.2, on the relevance and usefulness to the inference of x of an observation by an information source i. Clearly, a measurement z_i which occurs with unit probability contains the most information. This is in contrast to an observation z_i which does not change when the true state x changes. The usefulness or relevance of z_i can be measured by the extent to which its distribution changes with changes in x. Suppose the state changes from x to x', the joint distribution in each case is given by $p(\mathbf{z}, \mathbf{x})$ and $p(\mathbf{z}, \mathbf{x}')$ respectively. The difference in the distributions can be measured using distance measures based on entropy or Fisher information[7].

2.4.2 Probabilistic Information Update

Since the concept of entropy is the converse of information, a measure of information is given by negating entropy. Therefore, we can write, from the definition of entropy (Equation 2.54), that a measure of information is given by $E\{ln\, p(\mathbf{x})\}$. Whereupon the recursive Bayes update (Equation 2.14) can be written as an information update by taking expectations as follows

$$E\left\{ln\,[p(\mathbf{x} \mid \mathbf{Z}^k)]\right\} = E\left\{ln\,[p(\mathbf{x} \mid \mathbf{Z}^{k-1})]\right\} + E\left\{ln\left[\frac{p(\mathbf{z}(k) \mid \mathbf{x})}{p(\mathbf{z}(k) \mid \mathbf{Z}^{k-1})}\right]\right\}. \tag{2.65}$$

---

[7]For a discussion on the use of Fisher information for purposes of measuring distance between $p(\mathbf{z}, \mathbf{x})$ and $p(\mathbf{z}, \mathbf{x} + \delta\mathbf{x})$ see the discussion in Rao [182].

Informally, this can be interpreted as

$$posterior\ information\ =\ prior\ information$$
$$+\ observation\ information.$$

Similar information update equations can be written for the multi-information source systems based on the Independent Opinion Pool and the Independent Likelihood Pool i.e.

$$E\left\{ln\ [p(\mathbf{x}\mid\{\mathbf{Z}^k\})]\right\} = \sum_j E\left\{ln\ p(\mathbf{x}\mid\mathbf{Z}_j^k)\right\}, \qquad (2.66)$$

and

$$E\left\{ln\ [p(\mathbf{x}\mid\{\mathbf{Z}^k\})]\right\} = E\left\{ln\ [p(\mathbf{x}\mid\{\mathbf{Z}^{k-1}\})]\right\}$$
$$+ \sum_j E\left\{ln\ \left[\frac{p(\mathbf{z}_j(k)\mid\mathbf{x})}{p(\mathbf{z}_i(k)\mid\{\mathbf{Z}^{k-1}\})}\right]\right\}, \qquad (2.67)$$

respectively. A useful device is to simply write Bayes algorithm in terms of log-likelihoods as this also brings out the additive nature of the update.

2.4.3 Initialization, Priors and Robustness

According to Bayes Theorem (Equation 2.10), when no information is available subjectively about the state x, a prior PDF is required which contains no more information than is inherent in the likelihood in order to maintain objectivity. A non-informative prior is a PDF that contains no information about the state x. The subject of non-informative priors has deservedly received much attention [117][42] and the difficulties associated with them are generally regarded as one of the main short-comings of Bayesian analysis. Methods do exist for approximating non-informative priors. These are largely based on equating imprecision with ignorance, an example being the use of the *Maximum entropy principle*. Appendix A.2 discusses some of the approaches used to obtain non-informative priors. In a system where information update is recursive, crude methods which may be eschewed by Bayesian purists such as Maximum entropy priors and Fisher information priors, yield plausible results. Such results are possible with the *proviso* that the system has an adequately modelled likelihood function (observation model).

For most inference and estimation problems normally encountered in sensing, the system is generally robust, in a Bayesian sense. Here Bayesian robustness refers to the robustness of the posterior with regards to changes in the prior in the light of likelihood information according to Equation 2.13. Bayesian robustness is assured by having an informative likelihood function. With regard to sensing systems, the informativeness of the likelihood function is assured by the following considerations:

1. Since sensors are chosen and modelled such that they provide information relevant to the inference of the state, the likelihood is therefore informative.

2. In problems where several possibly unrelated measurements are taken, data association and validation assure that observations are correctly associated with the right states and therefore informative.

In addition, robustness is assured by the choice of priors such as non-informative priors and maximum entropy priors which have no impact on the posterior.

2.5 Summary

The probabilistic model presented provides a consistent framework for confronting issues in decentralized sensing in particular, and multi-sensor systems in general, in terms of:

- Development of architectures based on the methods of combining probabilistic information discussed in Section 2.2.

- Estimation and inference of the true state when presented with probabilistic information from a multiplicity of sources.

- Distributed and decentralized management based on the group decision principles set out in Section 2.3.2, utilizing results and properties from Section 2.4 on information and its update.

In presenting this model, we have attempted to illustrate the completeness and self-sufficiency of the Bayesian probabilistic approach by demonstrating how different components of the general problem outlined in Chapter 1 can be considered within the Bayesian paradigm.

The remainder of the material presented in this book is developed from the elements presented here. A method of tackling the important practical consideration of data association has been indicated although this is not pursued to any great detail in this book.

While the framework presented above is not in itself limited to fully-connected systems, there are extensions required with regards to communication in non-fully connected systems. Grime and Durrant-Whyte [90] have shown that by considering information sets using Bayes Theorem, the additional communication issues can be resolved, following which the model proposed in this chapter can be applied. In general, Bayesian calculations are particularly difficult when considering group decisions and indeed Tsitsiklis and Athans [202] show that minimizations similar to the maximizations of Equations 2.45-2.47 are NP-hard for a distributed system. However, it is possible, as we shall show, that given certain assumptions, methods can be developed for decision making which are not computationally taxing.

2.6 Bibliographical Note

The material presented in this chapter has drawn and brought together material originally presented in a variety of related areas. The classic text by Ferguson [77] presents the probabilistic decision theory on which much of the work in this chapter is based. Other standard texts include those by Jeffreys [117] and Rohatgi [184] and more recently Papoulis [170]. The excellent text by Berger [23] is particularly useful as a general and comprehensive introduction to probability theory and decision theory and henceforth this chapter. It is worth pointing out that the Bayesian approach is not universally accepted, admittedly there are some philosophical and technical difficulties, however Berger presents an ineluctable defence for a Bayesian approach. (See Berger [23] for the criticisms and the defence.). The theory of sufficient statistics was defined by Fisher [80] and further developed by Lehmann [132]. The literature on estimation and statistical inference is extensive, standard texts include those by Rao [182], Gelb [86] and Scharf [188]. Much of the literature on information fusion or combining probabilities and decision theory, comes from papers in diverse areas ranging from economics to robotics. Early references in this area include, Nash [163], Harsanyi [102], Luce and Raiffa [139] and Raiffa [176] and more recently Bacharach [11, 12] and DeGroot [63]. The recent surveys by

2.6 Bibliographical Note

Weerahandi and Zidek [205, 206] and the discussions in Berger [23] describe the current state of affairs. We presented a geometric interpretation of uncertainty based on considering uncertainty ellipses. In this vein, the work by Nakamura [161] and also Lee [131] is interesting as it pursues this to its logical conclusion by developing methods for geometric data fusion which are equivalent to what is presented in this book. The development of information theory has been synonymous with the developments in communication and coding theory. Formalization of much of this work is first noted in the seminal paper by Shannon [191]. The text by Cover and Thomas [52] and the references found therein, provide an excellent introduction to the subject. The use of information concepts in inference and estimation is mentioned in the texts by Maybeck [149] and Catlin [42].

3 Architectures and Algorithms

Though this be madness, yet there is method in't
- Shakespeare (Hamlet)

There are two themes in this chapter; firstly, the architectures for data fusion and secondly, algorithms for inference and estimation in data fusion. Both the architectures and the algorithms are developed in a mutually consistent manner by employing the probabilistic model of Chapter 2. In this context, architectures refer to the way sensing agents are inter-connected in order to facilitate the fusion of information.

Two representative data fusion algorithms are presented. The first of these algorithms is the celebrated Kalman filter which has seemingly countless applications [203][14][37]. What is presented, however, is the information form of the Kalman filter for the various architectures culminating in that for the decentralized architecture. The Kalman filter and its equivalent, the information filter (inverse covariance form), are not new algorithms and have been discussed elsewhere. Here, the various filters are developed directly from the information update relation introduced in Chapter 2. Such a derivation is consistent with the architectures discussed, thereby making the architectures a logical extension of the algorithms and *vice versa*. The second algorithm is the discrete equivalent of the information filter, which updates beliefs in the classification of a discrete state. This algorithm is similar to that presented by Rao [178] and also described in Rao and Manyika [181]. However, in the probabilistic model of Chapter 2, this algorithm comes about as a result of considering the state in the probabilistic information update of Equation 2.65 as discrete.

Deriving architectures and algorithms in this mutually consistent way means that these no longer need to be considered as distinct entities but as extensions of each other.

3.1 A Progression of Data Fusion Architectures

The methods of information fusion introduced in Chapter 2, lead to the development of centralized, hierarchical and, distributed and decentralized architectures for data fusion. Each of these has its own operational and practical advantages. In fact the development of each one has been in response to the shortcomings of the one preceding. Prior to developing these architectures formally, it is useful to discuss some of the features of these architectures.

In a fully *centralized* architecture, the central processor is responsible for collecting measurements from sensor devices, processing and interpreting the information obtained. Figure 3.1 illustrates the concept. Several features of the architecture can be noted:

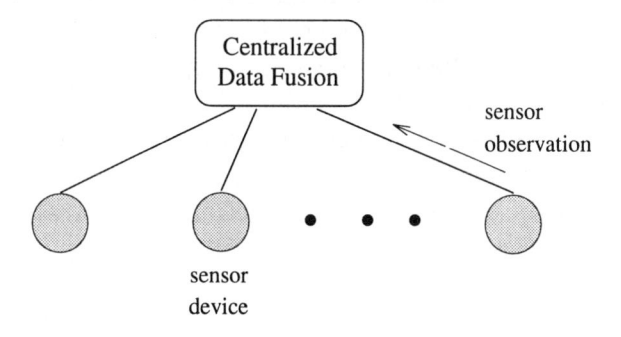

Figure 3.1: Centralized Architectures.

- Conceptually, the algorithms that can be employed are similar to those for single sensor systems and, as such, are relatively less complex.

- The central processor is able to make decisions based on the maximum possible information obtained directly from the environment. Resource management is, therefore, much easier because the central processor has the benefit of all the available information regarding the environment.

- Since all the information is available at the same site simultaneously, erroneous data from possibly faulty sensor devices can be easily identified and rejected. The architecture thus lends itself easily to outlier detection and consensus data validation algorithms.

- Since the central processor is fully aware of the information from each sensor and its activities, there is no possibility of tasks being done twice or information being fused more than once.

- Centralized architectures are inflexible to changes of application, sensor technology and sometimes expansion.

Although such centralized systems are clearly an improvement on single sensor systems, there several limitations: Firstly, centralised systems impose severe computational loads on the central processor. Depending on the complexity of the models being used, the model matrices can be quite large. Secondly, failure of the central processor would be disastrous since the operation of the entire system is dependent upon its well being. This is why, in practice, many such systems have a back-up processor for such eventualities. Such a contingency greatly increases cost and is not an efficient utilisation of computing power. And thirdly, the communication overhead can be considerable, given that all the sensor data has to be transferred to the central processor for processing.

Hierarchical architectures relieve the burdens on the central processor by allowing several levels of abstraction as illustrated in Figure 3.2. Typically, each sensor node (or local fusion center) maintains its

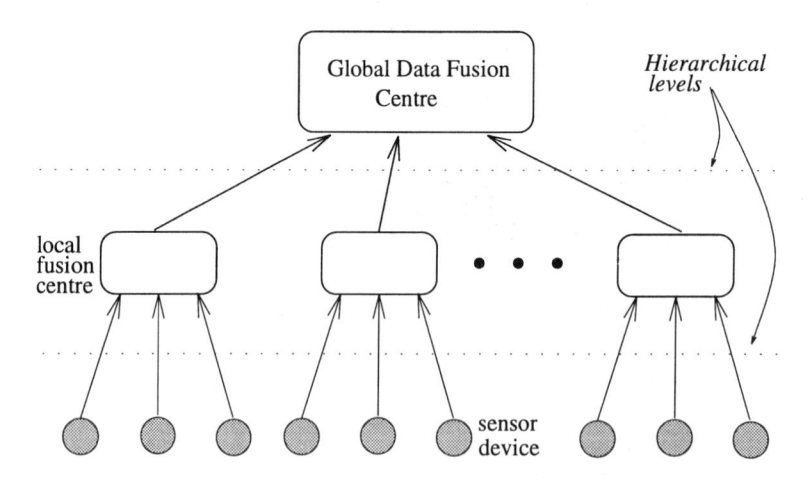

Figure 3.2: Hierarchical Architecture.

own track file (local estimates) based only on its own measurements (as would happen in a single sensor system). These local estimates are then sent to the central processor (or global fusion center) for fusion. Most

3.1 A Progression of Data Fusion Architectures

advanced systems in practical use employ this architecture or slight variations of it. The architecture, however, still retains a number of the disadvantages associated with the centralized architectures plus some additional ones:

- The architecture requires two types of algorithms, one for sensor level tracking and another for data fusion.

- Depending on the way fusion is implemented, there may be a requirement for two-way communications between the hierarchical levels, which would make the system vulnerable to communication bottlenecks.

Decentralized architectures offer ways of overcoming some of the problems associated with centralized and hierarchical architectures. In a decentralized architecture, each local sensor node makes observations and generates its own local estimate which is then communicated to other similar sensor nodes[1]. In return it receives local estimates from other nodes and then fuses these with its own to produce the global estimate. Such an architecture has all the advantages of a hierarchical one, plus increased reliability due to the absence of a central processor. Such an architecture is illustrated in Figure 3.3. Several features can be noted:

- Performance of the architecture is not dependent on any particular processor.

- Performance will deteriorate gradually as sensor nodes fail, since failure of a single node does not result in a failure of the whole system.

- Changes of sensor technology, etc, are easy to implement since such changes only affect a single node at a time, given the self-contained and autonomous nature of each node.

- The communication and algorithmic issues are decidedly more complex.

---

[1] We refer to each sensor capable of some processing in a multi-sensor system as a sensor node, node or simply sensor. Correspondingly, we use the terms local, nodal and partial, interchangeably to mean the result obtained using only information available locally at each sensor or node.

It is worth noting that a decentralized architecture has increased communication overheads compared to a centralized system but less than an equivalent hierarchy. This can be seen as follows, assuming that each nodes computes a local partial estimate vector of dimension n, the following can be observed. For a fully connected network with N nodes, a total of $(n^2 + n)(N - 1)$ numbers need to be communicated each cycle. This amount is more than that in an equivalent centralized system where each sensor device only communicates a sensor measurement. The total for an equivalent hierarchy is higher i.e. $3N(n^2 + n)$.

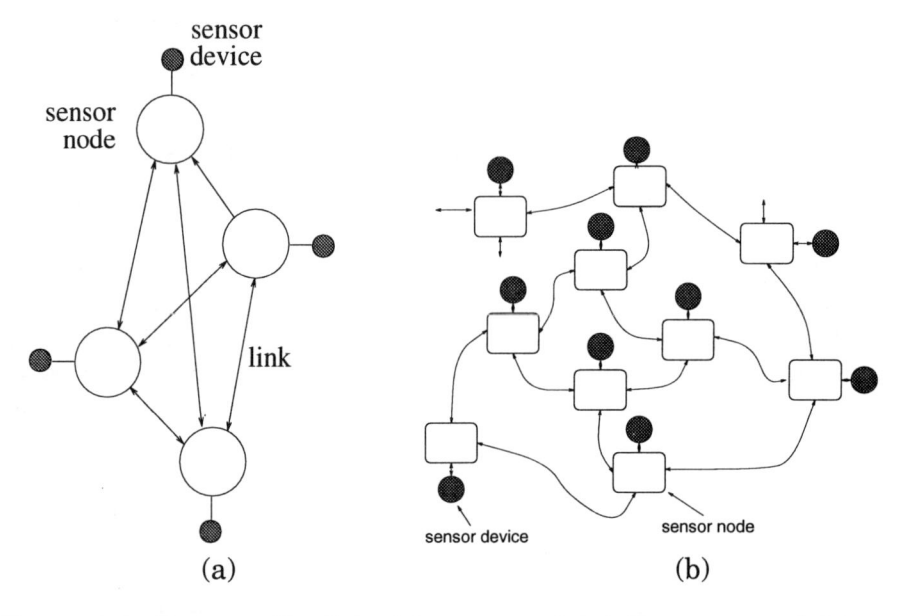

Figure 3.3: Decentralized Architectures. (a) fully connected (b) non-fully connected.

When configuring decentralized architectures, these can be either *fully connected* or *non-fully connected*. Fully connected architectures have each node connected directly to every node in the architecture. In a non-fully connected architecture, nodes are not necessarily connected to every other node. Figure 3.3 (a) illustrates a fully connected architecture and Figure 3.3 (b) illustrates a non-fully connected one. In a fully connected architecture consisting of N nodes, the number of communication links required is given by

$$links = \frac{(N-1)N}{2}. \tag{3.1}$$

3.1 A Progression of Data Fusion Architectures

In a non-fully connected architecture the number of links depends on the number available at each node. For example, the decentralized architectural hardware which we have developed based on Transputer architecture (see Appendix B.4), allows a maximum of three links per node (the fourth transputer link being dedicated to the sensor device). Therefore, with this implementation, the number of links in a non-fully connected architecture has a maximum i.e.

$$links_{max} = N + \frac{N}{2}, \qquad (3.2)$$

where the division $N/2$ is integer division. It can be seen from Equations 3.1 and 3.2 that the maximum number of links in a non-fully connected architecture (order $O(N)$), based on our implementation, is much less than that required by a similar size fully connected architecture (order $O(N^2)$). So it can be appreciated that non-fully connected architectures become increasingly pragmatic in large networks.

There are several difficulties to be encountered in the development and implementation of non-fully connected architectures. Non-fully connected systems give rise to additional complications in communication, routing and topological issues and, indeed, in the data fusion algorithms themselves. In fact these issues, with respect to non-fully connected systems, are active research areas but lie outside the scope of what is presented in this book. While, work has been done [148] in the area of fault-tolerant and verifiable communication in parallel implementations of fully connected decentralized systems, the equivalent for non-fully connected systems is yet to be done fully. The routing of information in non-fully connected systems can be addressed based on techniques that make use of spanning trees and topological maps. With regard to the additional complications to data fusion estimation, reference can be made to the recent work by Grime and Durrant-Whyte [90] on the correct propagation of estimation information in non-fully connected systems.

In this monograph, unless stated otherwise, references to decentralized architecture mean fully connected decentralized architectures. While most of the presentations in this book are with regards to decentralized architectures, we are not calling for whole-scale decentralization of *all* sensing activities. In fact there are applications where decentralization only serves to complicate matters.

3.2 Architectures and Information Update

The probabilistic information update (Equations 2.65-2.67) can be used as the basis for formally describing architectures for data fusion. A description is now given of how the information update is formulated for various architectures culminating in the update equations for a decentralized architecture.

3.2.1 A Generic Single Sensor Architecture

The information update for a single sensor system, written in terms of log-likelihoods, is

$$ln\ p(\mathbf{x} \mid \mathbf{Z}^k) = ln\ p(\mathbf{x} \mid \mathbf{Z}^{k-1}) + ln\ [\alpha\ p(\mathbf{z}(k) \mid \mathbf{x})]. \quad (3.3)$$

This equation describes a single Bayesian information source as discussed in Chapter 2. Such a single sensor system is illustrated in Figure 3.4. The system comprises the sensor device which provides observations $\mathbf{z}(k)$, a sensor observation model which describes the observation as likelihood information, i.e. the log of the probability distribution of the observation given a state $\mathbf{x}$. The likelihood information is combined with prior information i.e. the log of the distribution of the state given all the observation thus far, to give the posterior distribution of the state, given all the observations.

Multi-sensor systems are formed by combining elements of this single sensor system.

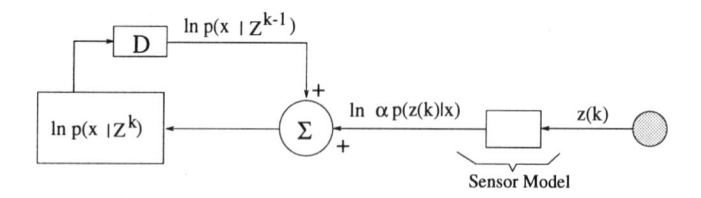

Figure 3.4: Single sensor information update.

3.2.2 Centralized and Hierarchical Architectures

Consider a centralized multi-sensor system with a central fusion processor. Each sensor node j makes local observations $\mathbf{z}_j(k)$ at time k

3.2 Architectures and Information Update

resulting in the set $\mathbf{Z}_j^k$ as defined in Equation 2.27. Each sensor communicates information to the central processor which fuses it with information from other sensors to construct a global posterior PDF. The computation of the global posterior depends on the method employed to combine probabilistic information, as we now discuss.

Independent Opinion Pool. If, according to the discussion in Chapter 2, it is known that each sensor node has subjective prior information then the Independent Opinion Pool of Section 2.2.1 can be used to combine information. A complete local posterior distribution $p(\mathbf{x} \mid \mathbf{Z}_i^k)$ is computed at each sensor i and then communicated to the central processor. In this case, the central processor then computes the global posterior as the summation

$$ln\, p(\mathbf{x} \mid \{\mathbf{Z}^k\}) = \sum_j \underbrace{ln\, p(\mathbf{x} \mid \mathbf{Z}_j^k)}_{\text{communicated}}, \tag{3.4}$$

where the local posterior is given by

$$ln\, p(\mathbf{x} \mid \mathbf{Z}_j^k) = \left[ln\, p(\mathbf{x} \mid \mathbf{Z}_j^{k-1}) + ln\, [\alpha_j\, p(\mathbf{z}_j(k) \mid \mathbf{x})] \right]. \tag{3.5}$$

Thus, the global posterior is simply a summation of the local posteriors from each sensor. An important consideration is that the prior used in Equation 3.5 is a local one which is updated as; $p(\mathbf{x} \mid \mathbf{Z}_i^k) \to p(\mathbf{x} \mid \mathbf{Z}_i^{k-1})$. This architecture is illustrated in Figure 3.5.

Independent Likelihood Pool. If no *local* subjective prior information is available, which is typically the case, the Independent Likelihood Pool is assumed. The central processor computes the global posterior using the following update rule

$$ln\, p(\mathbf{x} \mid \{\mathbf{Z}^k\}) = ln\, p(\mathbf{x} \mid \{\mathbf{Z}^{k-1}\}) + \sum_j ln\, [\alpha_j\, p(\mathbf{z}_j(k) \mid \mathbf{x})]. \tag{3.6}$$

The nature of the hierarchy determines the information communicated to the central processor by sensor i. This information may be in one of the following forms;

- raw sensor data $\mathbf{z}(k)$, or

- likelihoods $[\alpha_i\, p(\mathbf{z}_i(k) \mid \mathbf{x})]$, or alternatively, from considerations of sufficient statistics, $p(T(\mathbf{z}_i(k)) \mid \mathbf{x})$ (see Chapter 2).

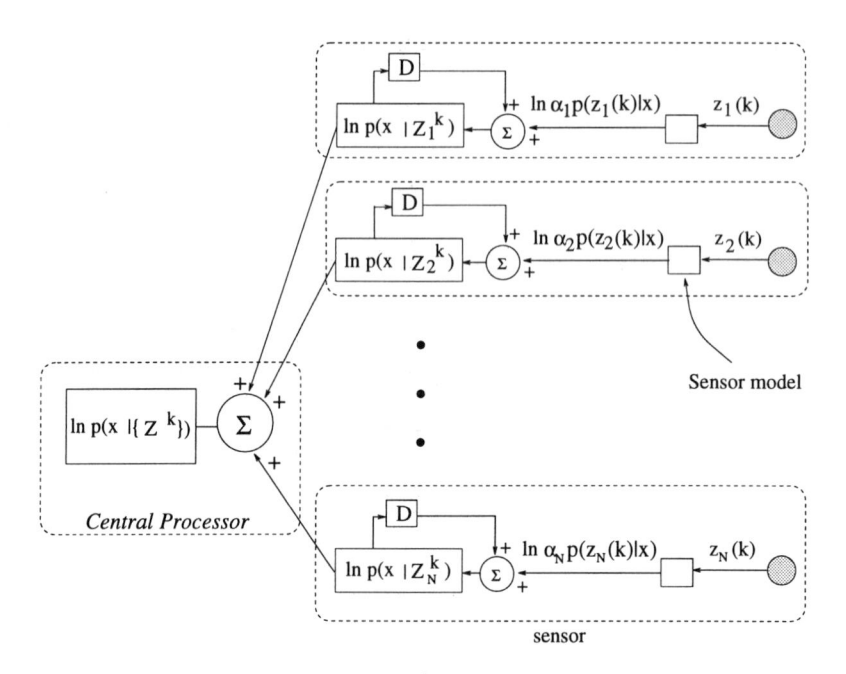

Figure 3.5: Centralized information update for architecture communicating local posterior information to central processor. This architecture is based on the Independent Opinion Pool.

The communication of raw sensor data requires the central processor to have knowledge of the models for each sensor, thereby, allowing it to compute the likelihoods $p(\mathbf{z}_i(k) \mid \mathbf{x})$ from the observations $\mathbf{z}_i(k)$, for all the sensors i. In this way, the global posterior $p(\mathbf{x} \mid \{\mathbf{Z}^k\})$ of Equation 3.6 is computed in its entirety at the central processor as illustrated in Figure 3.6. If each sensor has some computational capability then the likelihood term may be computed locally at each sensor and communicated. This has the advantage that the sensor model can be kept locally at each sensor. This architecture is illustrated in Figure 3.7. The global posterior is a summation of the prior global posterior and the likelihood information obtained at each sensor i.

Choosing between these hierarchies depends not only on the nature of the prior information, but also on the application. For example, a system comprising simple sensors such as temperature transducers is ideally suited to the hierarchy implied by Figure 3.6. The communication of likelihoods or local posteriors, shown in Figure 3.7 and Figure 3.5 respectively, means that the central processor need not have any

3.2 Architectures and Information Update

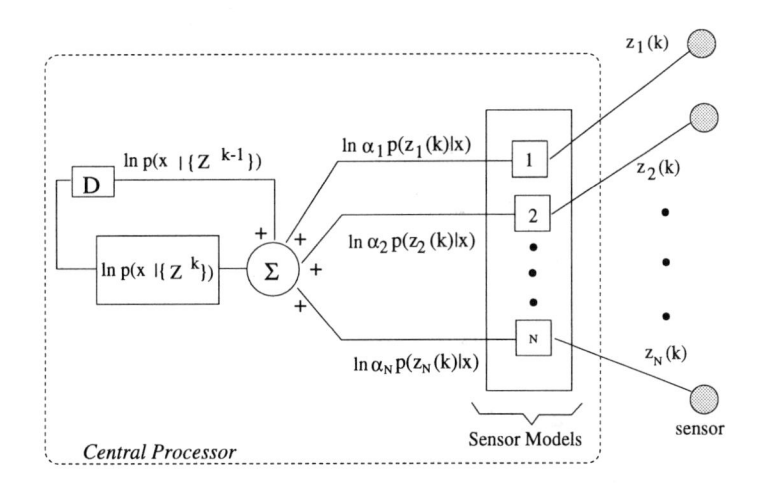

Figure 3.6: Centralized information update for architecture with sensor communicating actual observations.

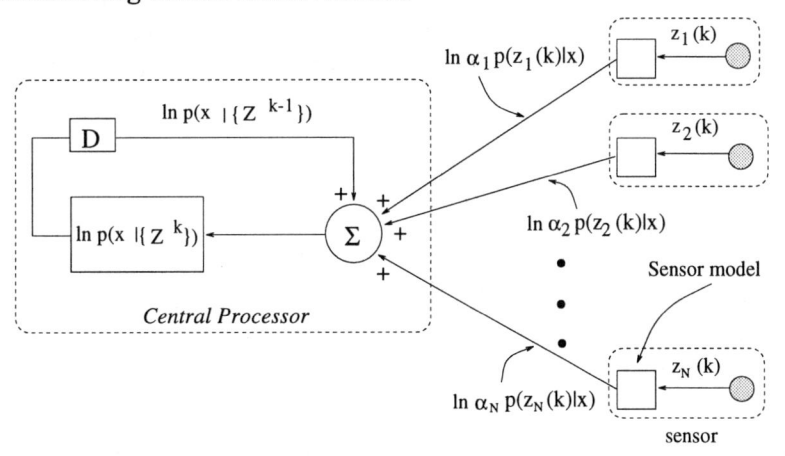

Figure 3.7: Centralized information update for architecture communicating likelihoods to central processor.

knowledge of the sensor models at each sensor node. This reduces the computational burden on the central processor. The hierarchies of Figure 3.7 and Figure 3.5 have the added advantage that a sensor can be changed or modified without the central processor needing to have any knowledge of it i.e. without requiring to update sensor models at the central processor. This makes the system modular and facilitates the inclusion of a wide range of sensor complexity in the system because any complex signal processing necessary for a given sensor can be realized locally. In Figure 3.5 initialization is done locally due to the fact that the prior is subjective at each sensor node. (See discussion in Section 2.2, Chapter 2).

The hierarchies in Figures 3.7 and 3.5 lead to the development of fully decentralized architectures.

3.2.3 Distributed and Decentralized Architectures

Each sensor node in a decentralized system, in addition to making observations, is required to obtain the global posterior after communicating with all the other sensors. For a fully connected system, the global posterior is expected to be *identical* at each sensor node. In order to formulate this architecture precisely, a local state vector is defined

$$\mathbf{x}_i \in \mathcal{X}, \quad \text{associated with sensor } i. \tag{3.7}$$

The decentralized architecture is a result of distributing the computations performed by the hierarchical architectures so that each sensor node is able to obtain the global posterior. As before, the decentralized architectures considered are those resulting from the Independent Opinion Pool and the Independent Likelihood Pool respectively:

Independent Opinion Pool. Consideration of the Independent Opinion Pool from Chapter 2, and the architecture suggested by the hierarchy of Equation 3.4 leads to a decentralized architecture. As before, the implicit assumption in the resulting architecture is that the observations conditioned on the state are independent *and* each sensor node has subjective prior information. A decentralized architecture of this form can be realized by placing the central processor of Equation 3.4 (see Figure 3.5) at each sensor node. In this way, each sensor node i communicates

3.2 Architectures and Information Update

a local posterior $p(\mathbf{x}_i \mid \mathbf{Z}_i^k)$ and obtains the global posterior according to

$$ln\, p(\mathbf{x}_i \mid \{\mathbf{Z}^k\}) = \sum_j \underbrace{ln\, p(\mathbf{x}_j \mid \mathbf{Z}_j^k)}_{\text{communicated}}, \qquad (3.8)$$

where the local posterior for each sensor node i is obtained as follows

$$ln\, p(\mathbf{x}_i \mid \mathbf{Z}_i^k) = ln\, p(\mathbf{x}_i \mid \mathbf{Z}_i^{k-1}) + ln\, [\alpha_i\, p(\mathbf{z}_i(k) \mid \mathbf{x}_i)]. \qquad (3.9)$$

A disadvantage of this architecture is that a given sensor node only has access to partial posterior information at other nodes (this resulting from the inter-nodal communication), and does not have direct access to the observed likelihood information other than its own. Therefore, sensor management approaches which require access to *all* the observed likelihood information in the system, cannot be implemented at each node without additional communication overheads.

Independent Likelihood Pool. In a decentralized system based on the Independent Likelihood Pool, communication is in the form of likelihood information[2]. At each sensor node all the likelihoods are fused to give the global posterior. This amounts to replicating the central processor function of Equation 3.6 at each sensor node. And as a result, he global posterior at sensor node i is obtained using the following log update relationship

$$ln\, p(\mathbf{x}_i \mid \{\mathbf{Z}^k\}) = ln\, p(\mathbf{x}_i \mid \{\mathbf{Z}^{k-1}\}) + \sum_j \underbrace{ln\, [\alpha_j\, p(\mathbf{z}_j(k) \mid \mathbf{x}_j)]}_{\text{communicated}}. \qquad (3.10)$$

This is illustrated for each sensor node in Figure 3.8. In addition, each sensor node i can also compute a *local partial posterior* based only on local observation information and a global prior, i.e.

$$ln\, p(\mathbf{x}_i \mid \{\mathbf{Z}^{k-1}\} \cup \mathbf{z}_i(k)) = ln\, p(\mathbf{x}_i \mid \{\mathbf{Z}^{k-1}\}) + ln\, [\alpha_i\, p(\mathbf{z}_i(k) \mid \mathbf{x}_i)]. \qquad (3.11)$$

This partial posterior has very important consequences because of what it represents, that is, a summation of information known globally before time k and the information that sensor i contributes at time k. This

---

[2] An important consequence of architectures where likelihoods are communicated is that a reduced order state can be used locally at each information source. This has been described in the work by Alouani [6] following on from the work described in [197][48][208].

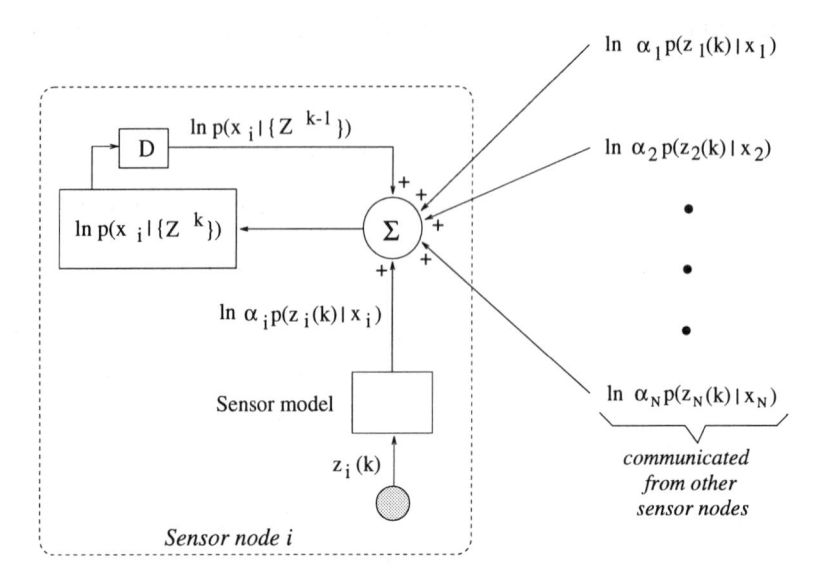

Figure 3.8: Decentralized information update at sensor node i. Sensor i communicates its likelihood to all the other sensors and in turn receives likelihoods from all the other sensors as shown.

is invaluable if it is required to evaluate the significance of sensor i's observation, its impact on and contribution to the global posterior etc. This is exploited in the development of sensor management strategies in Chapter 4.

3.2.4 Remarks

All the multi-sensor architectures above are equivalent for fully connected systems but differ in the way they combine probabilistic information, as discussed in Chapter 2, Section 2.2. The decentralized systems are more pragmatic in terms of robustness, performance and reliability. However, an important distinction should be reiterated, which is that architectures based on the Independent Opinion Pool can only be used when there is subjective prior information available locally, a rare condition in practice (see discussion in Chapter 2).

Assuming that each sensor node has subjective prior information, an interesting point to note about architectures based on the Independent Opinion Pool (Equations 3.4 and 3.8) is that they allow the implementation of hybrid algorithms as follows; each sensor can locally compute its partial posterior using any method for generating a posterior locally and then the global result is obtained by summing up the partial posterior information from all the nodes.

3.3 Multi-Sensor Continuous State Estimation

Ultimately, a data fusion system is required to estimate or infer the state vector which describes the state of nature, from the information available in the form of sensor measurements. The inference of a continuous state, using the classical methods introduced in Chapter 2, is now considered. The discussion then proceeds to consider the estimation of continuous states in a multi-sensor system.

The Kalman filter is an algorithm for recursively estimating the state of nature given a set of uncertain observations. The filter produces estimates of the state that are optimal in a statistical sense. The Kalman filter (KF) has been derived using a variety of approaches such as those based on the Gauss-Markov theorem as shown by Scharf [188] making assumptions on the joint distributions of the state and observations [149]. Other approaches make no assumptions on the joint distributions but apply orthogonality results to give least squares estimates [16]. Such approaches have been used to derive the standard formulation of the Kalman filter.

An alternative statement of the Kalman filter has been presented in the form of the *information filter* also called the inverse covariance form. Here the term "information" is used in the Fisher sense (i.e. inverse covariance matrix) and is directly related to the Bayesian probabilistic information. Formulations of the information form of the Kalman Filter have been presented by Maybeck [149], Grime and Durrant-Whyte [91] starting from the usual formulation of the Kalman filter. This information form of the Kalman filter has been shown to have some computational advantages especially when the state vector is of a greater dimension than the observation vector. In addition, the information filter overcomes some of the difficulties associated with initialization. Quite apart from the computational and practical advantages, the information filter is presented here because it can be stated and indeed developed in a manner which is consistent with the information update theme in this book. Therefore, what is presented here is an alternative *direct* derivation of the information form of the Kalman filter starting from a consideration of probabilistic information and its update.

3.3.1 The Information Filter

State Transition

The underlying state of nature is often continuous and varies with time. Such a state of nature can be described by an n-dimensional state vector $\mathbf{x}$, such that, $\mathbf{x} \in \mathcal{X}$ and $\mathcal{X} \subseteq \Re^n$. The dynamics of the state can be described using a state space representation [149]. For a linear system, this is written in the form of a first order vector differential equation i.e.

$$\dot{\mathbf{x}}_t = \mathbf{A}_t \mathbf{x}_t + \mathbf{w}_t, \tag{3.12}$$

where $\mathbf{x}_t$ is the state vector at time t, $\mathbf{A}$ is a known $n \times n$ matrix and $\mathbf{w}_t$ is the noise associated with the process.

In practical systems consisting of digital devices, sampling at discrete times is employed to obtain measurements and describe the instantaneous state of the system. The various components of the system at discrete sampling times are denoted by the index t_k. Therefore, Equation 3.12 can be written in discrete-time [16] as

$$\mathbf{x}_{t_k} = \mathbf{F}_{t_k, t_{k-1}} \mathbf{x}_{t_{k-1}} + \mathbf{w}_{t_k}, \tag{3.13}$$

where $\mathbf{F}$ is the *state transition matrix* describing the transition of the state from time-step to time-step and is given by

$$\mathbf{F}_{t_k, t_{k-1}} = \exp\left\{\mathbf{A}_{t_k, -t_{k-1}}\right\} = \mathbf{F}(k). \tag{3.14}$$

The noise $\mathbf{w}_{t_k}$ is given by

$$\mathbf{w}_{t_k} = \int_{t_{k-1}}^{t_k} \exp\left\{\mathbf{A}_{t_k, -\tau}\right\} \mathbf{w}_\tau \, d\tau = \mathbf{w}(k). \tag{3.15}$$

The accuracy of modelling the noise in the way can be improved by taking the mean value of $\mathbf{w}_t$ over the sampling interval. Strictly, Equations 3.14 and 3.15 hold for a sampled time-invariant system. Equation 3.13 can be conveniently written by dropping the time index since the sample time t is usually constant. Hence, in discrete-time notation we write a linear state transition of the form

$$\mathbf{x}(k) = \mathbf{F}(k)\mathbf{x}(k-1) + \mathbf{w}(k), \tag{3.16}$$

where $\mathbf{F}(k)$ is the *state transition matrix* and $\mathbf{w}(k)$ is the state transition noise which is assumed to be zero-mean Gaussian. Thus, the transition

3.3 Multi-Sensor Continuous State Estimation

noise is uncorrelated in time and satisfies the conditions

$$E\{\mathbf{w}(k)\} = 0,$$
$$E\{\mathbf{w}(k)\mathbf{w}(j)^T\} = \mathbf{Q}(k)\delta_{k,j}, \quad \forall k,j \quad (3.17)$$

where δ is the Kronecker delta operator and the covariance matrix $\mathbf{Q}(k)$ is given by

$$\mathbf{Q}(k) = \int_{t_{k-1}}^{t_k} \exp\{\mathbf{A}_{t_k,-\tau}\} \mathbf{w}(\tau) \exp\{\mathbf{A}_{t_k,-\tau}^T\} d\tau. \quad (3.18)$$

$\mathbf{Q}(k)$ is positive and semi-definite. It can be noted that Equation 3.16 represents a time varying system.

Observation

Observations are made which contain information about the state of nature according to Equation 2.2. For a continuous system, the observations make up an m-dimensional observation vector $\mathbf{z}$, such that, $\mathbf{z}$ in $\mathcal{Z}$ and $\mathcal{Z} \subseteq \Re^m$. A linear form of the observation model is assumed and this is written

$$\mathbf{z}_t = \mathbf{H}_t \mathbf{x}_t + \mathbf{v}_t, \quad (3.19)$$

where $\mathbf{H}$ is a $m \times n$ matrix which relates observations $\mathbf{z}$ to the state $\mathbf{x}$, and $\mathbf{v}$ is the noise associated with the observation. In practical systems where measurements are taken at sampled intervals, Equation 3.19 can be written in discrete-time as

$$\mathbf{z}_{t_k} = \mathbf{H}_{t_k} \mathbf{x}_{t_k} + \mathbf{v}_{t_k}. \quad (3.20)$$

Dropping the time index as before[3], we can write that

$$\mathbf{z}(k) = \mathbf{H}(k)\mathbf{x}(k) + \mathbf{v}(k), \quad (3.21)$$

where $\mathbf{H}(k)$ is an *observation matrix* and $\mathbf{v}(k)$, the observation noise, is zero mean Gaussian with covariance matrix $\mathbf{R}(k)$. Since the noise $\mathbf{v}(k)$ is a process which is uncorrelated in time, it satisfies

$$E\{\mathbf{v}(k)\} = 0,$$
$$E\{\mathbf{v}(k)\mathbf{v}(j)^T\} = \mathbf{R}(k)\delta_{k,j}, \quad \forall k,j. \quad (3.22)$$

---
[3] Hereafter, our notation shall be in terms of the discrete-time index k.

The observation noise covariance matrix $\mathbf{R}(k)$ is also positive and semi-definite and the off-diagonal terms are cross-correlation terms.

An additional assumption made is that the observation noise $\mathbf{v}(k)$ and the state transition noise $\mathbf{w}(k)$ are uncorrelated, that is

$$E\{\mathbf{v}(k)\mathbf{w}(j)\} = 0, \quad \forall k,j. \qquad (3.23)$$

Information Update and Estimation

The information update relationship can be written in an equivalent form using log-likelihoods as

$$l \triangleq ln\, p(\mathbf{x}(k) \mid \mathbf{Z}^k) = ln\, p(\mathbf{x}(k) \mid \mathbf{Z}^{k-1}) + ln[\alpha\, p(\mathbf{z}(k) \mid \mathbf{x}(k))]. \qquad (3.24)$$

It is assumed that all the PDFs in Equation 3.24 are Gaussian such that for a distribution $p(\mathbf{x})$,

$$p(\mathbf{x}) = N(\bar{\mathbf{x}}, \mathbf{P}), \quad \text{where } \bar{\mathbf{x}} = \text{mean, and } \mathbf{P} = \text{covariance.} \qquad (3.25)$$

The state can now be estimated recursively in order to obtain an MMSE estimate for $\mathbf{x}(k)$, given all the observations up to time-step k. This estimate is denoted $\hat{\mathbf{x}}(k \mid k)$. From the posterior in Equation 3.24, an MMSE estimate is obtained using Equations 2.21 and 2.22 as

$$\hat{\mathbf{x}}(k \mid k) = E\left\{\mathbf{x}(k) \mid \mathbf{Z}^k\right\}, \qquad (3.26)$$

with covariance

$$\mathbf{P}(k \mid k) = E\left\{(\mathbf{x}(k) - \hat{\mathbf{x}}(k \mid k))(\mathbf{x}(k) - \hat{\mathbf{x}}(k \mid k))^T \mid \mathbf{Z}^k\right\}. \qquad (3.27)$$

Corresponding to this estimate, a transformed state vector is defined which is called the *information state vector*. The information state vector at time-step $(j \mid l)$ is given by

$$\hat{\mathbf{y}}(j \mid l) \triangleq \mathbf{P}^{-1}(j \mid l)\hat{\mathbf{x}}(j \mid l). \qquad (3.28)$$

Associated with the information state vector, is the *information matrix*, which is given by the inverse covariance $\mathbf{P}^{-1}(j \mid l)$. It is important to note that $\mathbf{P}^{-1}(j \mid l)$ is really the covariance of the information state vector $\hat{\mathbf{y}}(j \mid l)$, i.e. $\mathbf{P}^{-1}(j \mid l)\hat{\mathbf{x}}(j \mid l)$.

These information estimates can be obtained based on the above assumptions by; (i) considering the probabilistic information update

3.3 Multi-Sensor Continuous State Estimation

relationship l of Equation 3.24, and substituting in the appropriate Gaussian PDFs, (ii) evaluating $\nabla_{\mathbf{x}} l$ and (iii) $\nabla_{\mathbf{x}} \nabla_{\mathbf{x}}^T l$. Following these steps, the information filter is derived.

Substituting for the Gaussian PDFs in l, and evaluating $\nabla_{\mathbf{x}} l$ making use of the result $\nabla_{\mathbf{x}}(\mathbf{y}^T \mathbf{A} \mathbf{y}) = 2(\nabla_{\mathbf{x}} \mathbf{y}^T)\mathbf{A}\mathbf{y}$, where $\mathbf{A}$ is symmetric and independent of $\mathbf{x}$, the following is obtained; for the term corresponding to the posterior

$$\nabla_{\mathbf{x}}\left[c_1 - \frac{1}{2}(\mathbf{x}(k) - \hat{\mathbf{x}}(k \mid k))^T \mathbf{P}^{-1}(k \mid k)(\mathbf{x}(k) - \hat{\mathbf{x}}(k \mid k))\right]$$
$$= -\left[\nabla_{\mathbf{x}}(\mathbf{x}(k) - \hat{\mathbf{x}}(k \mid k))\mathbf{P}^{-1}(k \mid k)(\mathbf{x}(k) - \hat{\mathbf{x}}(k \mid k))\right]$$
$$= -\mathbf{P}^{-1}(k \mid k)(\mathbf{x}(k) - \hat{\mathbf{x}}(k \mid k)), \quad (3.29)$$

and for the term corresponding to the prior

$$\nabla_{\mathbf{x}}\left[c_2 - \frac{1}{2}(\mathbf{x}(k) - \hat{\mathbf{x}}(k \mid k-1))^T \mathbf{P}^{-1}(k \mid k-1)(\mathbf{x}(k) - \hat{\mathbf{x}}(k \mid k-1))\right]$$
$$= -\mathbf{P}^{-1}(k \mid k-1)(\mathbf{x}(k) - \hat{\mathbf{x}}(k \mid k-1)), \quad (3.30)$$

and, similarly, for that corresponding to the likelihood

$$\nabla_{\mathbf{x}}\left[c_3 - \frac{1}{2}(\mathbf{z}(k) - \mathbf{H}(k)\mathbf{x}(k))^T \mathbf{R}^{-1}(k)(\mathbf{z}(k) - \mathbf{H}(k)\mathbf{x}(k))\right]$$
$$= -\left[\nabla_{\mathbf{x}}(\mathbf{z}(k) - \mathbf{H}(k)\mathbf{x}(k))^T\right]\mathbf{R}^{-1}(k)(\mathbf{z}(k) - \mathbf{H}(k)\mathbf{x}(k))$$
$$= \mathbf{H}(k)^T \mathbf{R}^{-1}(k)(\mathbf{z}(k) - \mathbf{H}(k)\mathbf{x}(k)), \quad (3.31)$$

where c_1, c_2 and c_3 are constants independent of $\mathbf{x}(k)$. Substituting Equations 3.29-3.31 in the expression for $\nabla_{\mathbf{x}} l$, where l is given by Equation 3.24, yields the result

$$\begin{aligned}-\mathbf{P}^{-1}(k \mid k)(\mathbf{x}(k) - \hat{\mathbf{x}}(k \mid k-1)) \\ = -\mathbf{P}^{-1}(k \mid k-1)(\mathbf{x}(k) - \hat{\mathbf{x}}(k \mid k)) \\ +\mathbf{H}(k)^T \mathbf{R}^{-1}(k)(\mathbf{z}(k) - \mathbf{H}(k)\mathbf{x}(k)).\end{aligned} \quad (3.32)$$

From the definition of Fisher information (Equations 2.61 and 2.62), $\nabla_{\mathbf{x}} \nabla_{\mathbf{x}}^T l$ gives the Fisher information matrix. Therefore, applying $\nabla_{\mathbf{x}}$ again to Equation 3.32 evaluates to

$$\mathbf{P}^{-1}(k \mid k) = \mathbf{P}^{-1}(k \mid k-1) + \mathbf{H}(k)^T \mathbf{R}^{-1}(k)\mathbf{H}(k). \quad (3.33)$$

Post-multiplying by $\hat{\mathbf{x}}(k \mid k)$ and adding to Equation 3.32 gives the following result

$$\mathbf{P}^{-1}(k \mid k)\hat{\mathbf{x}}(k \mid k) = \mathbf{P}^{-1}(k \mid k-1)\hat{\mathbf{x}}(k \mid k-1) + \mathbf{H}(k)^T \mathbf{R}^{-1}(k)\mathbf{z}(k).$$
(3.34)

Using the definition of the information state vector (Equation 3.28), Equation 3.34 can be written as

$$\hat{\mathbf{y}}(k \mid k) = \hat{\mathbf{y}}(k \mid k-1) + \mathbf{H}(k)^T \mathbf{R}^{-1}(k)\mathbf{z}(k), \quad (3.35)$$

which is the update equation for the information state vector. The predicted covariance $\mathbf{P}(k \mid k-1)$ is obtained from

$$\mathbf{P}(k \mid k-1) \triangleq E\left\{(\mathbf{x}(k) - \hat{\mathbf{x}}(k \mid k-1))(\mathbf{x}(k) - \hat{\mathbf{x}}(k \mid k-1))^T \mid \mathbf{Z}^{k-1}\right\}.$$
(3.36)

Since,

$$(\mathbf{x}(k) - \hat{\mathbf{x}}(k \mid k-1)) = \mathbf{F}(k)\left[(\mathbf{x}(k-1) - \hat{\mathbf{x}}(k-1 \mid k-1))\right] + \mathbf{w}(k),$$
(3.37)

the predicted covariance becomes

$$\mathbf{P}(k \mid k-1) = \mathbf{F}(k)\mathbf{P}(k-1 \mid k-1)\mathbf{F}(k)^T + \mathbf{Q}(k). \quad (3.38)$$

And so the predicted information state vector $\hat{\mathbf{y}}(k \mid k-1)$, is

$$\begin{aligned}\hat{\mathbf{y}}(k \mid k-1) &= \mathbf{P}^{-1}(k \mid k-1)\mathbf{F}(k)\hat{\mathbf{x}}(k-1 \mid k-1) \\ &= \mathbf{P}^{-1}(k \mid k-1)\mathbf{F}(k)\mathbf{P}(k-1 \mid k-1)\hat{\mathbf{y}}(k-1 \mid k-1).\end{aligned}$$
(3.39)

It will be noted that in the above formulation, the vector $\mathbf{H}(k)^T \mathbf{R}^{-1}(k)\mathbf{z}(k)$ is a sufficient statistic of $\mathbf{x}(k)$. Hence, from our discussion in Chapter 2 concerning sufficient statistics, this vector is a model of the likelihood $p(\mathbf{z}(k) \mid \mathbf{x}(k))$. That $\mathbf{H}(k)^T \mathbf{R}^{-1}(k)\mathbf{z}(k)$ is a sufficient statistic for $\mathbf{x}(k)$, can be seen by writing the Gaussian $[\alpha \, p(\mathbf{z}(k) \mid \mathbf{x})]$ in the form

$$\begin{aligned} p(\mathbf{z}(k) \mid \mathbf{x}) &\propto c\exp\left\{-\frac{1}{2}\mathbf{x}(k)^T \mathbf{H}(k)^T \mathbf{R}^{-1}(k)\mathbf{H}(k)\mathbf{x}(k)\right\} \\ &\quad \times \exp\left\{-\frac{1}{2}\mathbf{z}(k)^T \mathbf{R}^{-1}(k)\mathbf{z}(k)\right\} \\ &\quad \times \exp\left\{\mathbf{x}(k)^T \mathbf{H}(k)^T \mathbf{R}^{-1}(k)\mathbf{z}(k)\right\}, \end{aligned} \quad (3.40)$$

3.3 Multi-Sensor Continuous State Estimation

which by the factorization theorem[4] is the necessary condition for $\mathbf{H}(k)^T \mathbf{R}^{-1}(k)\mathbf{z}(k)$ to be sufficient for $\mathbf{x}(k)$. The statistic is normal with covariance $\mathbf{H}(k)^T \mathbf{R}^{-1}(k)\mathbf{H}(k)$. It has been shown in [188], that this sufficient statistic is unique (complete) and also minimal.

Non-linear Information Filter

Often the observation and transition equations are not linear and a non-linear information filter is required. Here we also use the same discrete-time notation which has already been introduced. For a non-linear system, the observation equation, based on Equation 2.2 and assuming additive observation noise, is written

$$\mathbf{z}(k) = \mathbf{h}[k, \mathbf{x}(k)] + \mathbf{v}(k), \qquad (3.41)$$

where $\mathbf{h}[\cdot]$ is the non-linear observation model transforming $\mathbf{x}(k)$ from the state space $\mathcal{X}$, to the observation space $\mathcal{Z}$. The additive observation noise is of the same form as in Equation 3.22. Similarly, the non-linear state transition equation can be written as

$$\mathbf{x}(k) = \mathbf{f}[k, \mathbf{x}(k-1)] + \mathbf{w}(k), \qquad (3.42)$$

where $\mathbf{f}[\cdot]$ is the non-linear state transition model describing the transition of the state from time-step to time-step as a non-linear function of the state. The non-linear filter which can be developed from this is the information form of the Extended Kalman Filter (EKF) [16].

The derivation proceeds as outlined earlier by first substituting the appropriate Gaussians into Equation 3.24. However, the Gaussian distribution corresponding to the likelihood $ln[\alpha \, p(\mathbf{z}(k)) \mid \mathbf{x}(k)]$ is different from that in Equation 3.31 and reflects the non-linearities in Equation 3.41 i.e.

$$ln[\alpha \, p(\mathbf{z}(k) \mid \mathbf{x}(k))] =$$
$$c - \frac{1}{2}(\mathbf{z}(k) - \mathbf{h}[k, \mathbf{x}(k)])^T \mathbf{R}^{-1}(k)(\mathbf{z}(k) - \mathbf{h}[k, \mathbf{x}(k)]). \qquad (3.43)$$

Proceeding as before, taking (i) $\nabla_\mathbf{x} l$ and (ii) $\nabla_\mathbf{x} \nabla_\mathbf{x}^T l$ and making the appropriate substitutions gives the result that the information state vector is updated according to

$$\hat{\mathbf{y}}(k \mid k) = \hat{\mathbf{y}}(k \mid k-1) + \nabla_\mathbf{x} \mathbf{h}[k, \hat{\mathbf{x}}(k \mid k)]^T \mathbf{R}^{-1}(k) \mathbf{z}'(k), \qquad (3.44)$$

---

[4] A vector $\mathbf{t}$ is sufficient for the vector $\mathbf{x}$ iff $p(\mathbf{z} \mid \mathbf{x}) = p(\mathbf{z})p(\mathbf{t} \mid \mathbf{x})$, and the distribution $p(\mathbf{z} \mid \mathbf{x})$ is normal [188].

where $\mathbf{z}'(k)$ is given by[5]

$$\mathbf{z}'(k) = \mathbf{z}(k) - (\mathbf{h}[k, \hat{\mathbf{x}}(k \mid k-1)] - \nabla_{\mathbf{x}} \mathbf{h}[k, \hat{\mathbf{x}}(k \mid k-1)] \hat{\mathbf{x}}(k \mid k-1)). \tag{3.45}$$

The inverse covariance (information matrix) is updated according to

$$\begin{aligned} \mathbf{P}^{-1}(k \mid k) &= \mathbf{P}^{-1}(k \mid k-1) \\ &+ \nabla_{\mathbf{x}} \mathbf{h}[k, \hat{\mathbf{x}}(k \mid k)]^T \mathbf{R}^{-1}(k) \nabla_{\mathbf{x}} \mathbf{h}[k, \hat{\mathbf{x}}(k \mid k)]. \end{aligned} \tag{3.46}$$

It can be shown that

$$\begin{aligned} (\mathbf{x}(k) - \hat{\mathbf{x}}(k \mid k-1)) &= \\ \nabla_{\mathbf{x}} \mathbf{f}[k, \hat{\mathbf{x}}(k-1 \mid k-1)] & (\mathbf{x}(k-1) - \hat{\mathbf{x}}(k-1 \mid k-1)) \\ + \underbrace{\cdots}_{\text{higher order terms}} & + \mathbf{w}(k), \end{aligned} \tag{3.47}$$

based on a linearization about the predicted state $\hat{\mathbf{x}}(k \mid k-1)$, where the higher order terms can be obtained from a Taylor series expansion [16]. By ignoring the higher order terms, the predicted covariance is given by

$$\begin{aligned} \mathbf{P}(k \mid k-1) &= [\nabla_{\mathbf{x}} \mathbf{f}[k, \hat{\mathbf{x}}(k-1 \mid k-1)] \mathbf{P}(k-1 \mid k-1) \\ &\times \nabla_{\mathbf{x}} \mathbf{f}[k, \hat{\mathbf{x}}(k-1 \mid k-1)]^T \Big] + \mathbf{Q}(k). \end{aligned} \tag{3.48}$$

Consequently, the predicted information state vector is obtained as

$$\hat{\mathbf{y}}(k \mid k-1) = \mathbf{P}^{-1}(k \mid k-1) \mathbf{f}[k, \hat{\mathbf{x}}(k-1 \mid k-1)]. \tag{3.49}$$

It must be noted that to evaluate $\nabla_{\mathbf{x}} \mathbf{h}[k, \hat{\mathbf{x}}(k \mid k-1)]$, the predicted state $\hat{\mathbf{x}}(k \mid k-1)$ is obtained from

$$\hat{\mathbf{x}}(k \mid k-1) = \mathbf{P}(k \mid k-1) \hat{\mathbf{y}}(k \mid k-1). \tag{3.50}$$

In [159], an alternative derivation of the non-linear information filter is presented.

In the formulation of the non-linear information filter, the following equivalences with the linear information filter are apparent

$$\nabla_{\mathbf{x}} \mathbf{h}[k, \hat{\mathbf{x}}(k \mid k)] \equiv \mathbf{H}(k) \quad \text{and} \quad \nabla_{\mathbf{x}} \mathbf{f}[k, \hat{\mathbf{x}}(k \mid k)] \equiv \mathbf{F}(k). \tag{3.51}$$

---

[5]The authors are grateful to A. Mutambara for showing this full expansion of $\mathbf{z}'$. With the derivation done naively, the result that $\mathbf{z}' = \mathbf{z}$ can be obtained.

3.3 Multi-Sensor Continuous State Estimation

And so, it follows that making the substitutions $\mathbf{H}(k)$ for $\nabla_\mathbf{x} \mathbf{h}[k, \hat{\mathbf{x}}(k \mid k)]$ and $\mathbf{F}(k)$ for $\nabla_\mathbf{x} \mathbf{f}[k, \hat{\mathbf{x}}(k \mid k)]$, makes the formulation of the first order non-linear information filter equivalent to that for the linear information filter (with the exception of Equation 3.49). The following box is a summary of the information filter.

Information Filter Equations

The information state vector is updated according to

$$\hat{\mathbf{y}}(k \mid k) = \hat{\mathbf{y}}(k \mid k-1) + \mathbf{H}(k)^T \mathbf{R}^{-1}(k) \mathbf{z}'(k), \qquad (3.52)$$

where $\mathbf{z}'(k) = \mathbf{z}(k)$ for the linear filter filter and is given by Equation 3.45 for the non-linear filter. The inverse covariance (information matrix) according to

$$\mathbf{P}^{-1}(k \mid k) = \mathbf{P}^{-1}(k \mid k-1) + \mathbf{H}(k)^T \mathbf{R}^{-1}(k) \mathbf{H}(k). \qquad (3.53)$$

The inverse covariance and information state vector are predicted as

$$\mathbf{P}(k \mid k-1) = \mathbf{F}(k) \mathbf{P}(k-1 \mid k-1) \mathbf{F}(k)^T + \mathbf{Q}(k), \qquad (3.54)$$

$$\hat{\mathbf{y}}(k \mid k-1) = \mathbf{P}^{-1}(k \mid k-1) \mathbf{F}(k) \mathbf{P}(k-1 \mid k-1) \hat{\mathbf{y}}(k-1 \mid k-1), \qquad (3.55)$$

respectively. The estimated state may be obtained from

$$\hat{\mathbf{x}}(k \mid k) = \mathbf{P}(k \mid k) \hat{\mathbf{y}}(k \mid k). \qquad (3.56)$$

The information filter algorithm which we have described is for a single sensor system such as that in Figure 3.4. However, the sensing systems we are interested in consist of several sensors in any of the architectural configuration discussed in Section 3.2. Hence, we concern ourselves with the estimation of continuous states in multi-sensor systems. The information filter can indeed be extended into algorithms which can be applied to multi-sensor systems.

3.3.2 Extension to Hierarchical Estimation

Hierarchical estimation has previously been discussed based on the usual formulation of the Kalman filter [48][91]. What is presented here outlines methods of deriving the appropriate form of the information filter from the information update, given an architectural paradigm. As this section is only an outline, scant details are given.

The hierarchical architecture can be treated as a single system in which the observation vector consists of partitions each contributed by each sensor. The following definitions are made with respect to the observation partitioning

$$\begin{aligned}
\text{observations } \mathbf{z}(k) &= [\mathbf{z}_1(k)^T, \mathbf{z}_2(k)^T, \ldots, \mathbf{z}_N(k)^T]^T \\
\text{observation matrix } \mathbf{H}(k) &= [\mathbf{H}_1(k)^T, \mathbf{H}_2(k)^T, \ldots, \mathbf{H}_N(k)^T]^T \\
\text{observation noise covariance } \mathbf{R}(k) &= diag[\mathbf{R}_1(k), \mathbf{R}_2(k), \ldots, \mathbf{R}_N(k)],
\end{aligned} \tag{3.57}$$

where the observation partitions are assumed to be uncorrelated. A centralized or hierarchical architecture is described by Equations 3.6 and 3.4 and illustrated in Figures 3.6 and 3.5. The filters resulting from these architectures correspond to the Independent Opinion Pool and Independent Likelihood Pool respectively. These are outlined in turn.

Independent Opinion Pool. For an architecture defined by Equation 3.4, where each sensor i communicates a local posterior $p(\mathbf{x} \mid \mathbf{Z}_i^k)$, the global posterior is computed from

$$ln\, p(\mathbf{x} \mid \{\mathbf{Z}^k\}) = \sum_j \underbrace{ln\, p(\mathbf{x} \mid \mathbf{Z}_j^k)}_{\text{communicated}},$$

where for each sensor node i, the local posterior is given by

$$ln\, p(\mathbf{x} \mid \mathbf{Z}_i^k) = ln\, p(\mathbf{x} \mid \mathbf{Z}_i^{k-1}) + ln\, [\alpha_i\, p(\mathbf{z}_i(k) \mid \mathbf{x})], \tag{3.58}$$

provided that each sensor has available subjective prior information. By following the steps outlined in Section 3.3.1, the corresponding filter for this architecture can be derived. The global information state vector is obtained as

$$\hat{\mathbf{y}}(k \mid k) = \sum_j \hat{\mathbf{y}}_j(k \mid k), \tag{3.59}$$

3.3 Multi-Sensor Continuous State Estimation

and the information matrix

$$\mathbf{P}^{-1}(k \mid k) = \sum_j \mathbf{P}_j^{-1}(k \mid k), \tag{3.60}$$

where the local information state vectors are given by

$$\hat{\mathbf{y}}_i(k \mid k) = \hat{\mathbf{y}}_i(k \mid k-1) + \mathbf{H}_i^T(k)\mathbf{R}_i^{-1}(k)\mathbf{z}_i(k), \tag{3.61}$$

and the local information matrices by

$$\mathbf{P}_i^{-1}(k \mid k) = \mathbf{P}_i^{-1}(k \mid k-1) + \mathbf{H}_i^T(k)\mathbf{R}_i^{-1}(k)\mathbf{H}_i(k). \tag{3.62}$$

The predictions are local versions of those in Section 3.3.1, that is

$$\mathbf{P}_i(k \mid k-1) = \mathbf{F}_i(k)\mathbf{P}_i(k-1 \mid k-1)\mathbf{F}_i(k)^T + \mathbf{Q}_i(k), \tag{3.63}$$

for the covariance and

$$\hat{\mathbf{y}}_i(k \mid k-1) = \mathbf{P}_i^{-1}(k \mid k-1)\mathbf{F}_i(k)\mathbf{P}_i(k-1 \mid k-1)\hat{\mathbf{y}}_i(k-1 \mid k-1), \tag{3.64}$$

for the information state vector. Since this filter is based on the Independent Opinion Pool, the prior information caveat is reiterated, i.e. that the prior information at each node must be locally subjective (see Chapter 2). In practice, this condition is rarely met and consequently this filter is not used often.

Independent Likelihood Pool. For the hierarchy described by Equation 3.6, where each sensor node communicates likelihood information, the log-likelihood l, is given by

$$ln \, p(\mathbf{x} \mid \{\mathbf{Z}^k\}) = ln \, p(\mathbf{x} \mid \{\mathbf{Z}^{k-1}\}) + \sum_j \underbrace{ln \, [\alpha_j \, p(\mathbf{z}_j(k) \mid \mathbf{x})]}_{\text{communicated}}.$$

Each sensor j makes an observation according to

$$\mathbf{z}_j(k) = \mathbf{H}_j(k)\mathbf{x}(k) + \mathbf{v}_j(k), \quad \forall j \in \mathcal{N}, \tag{3.65}$$

where the observation noise $\mathbf{v}_j(k)$ is under the same assumptions as in Equation 3.22.

In the information filter derived in Section 3.3.1, information from the observation is $\mathbf{H}(k)^T\mathbf{R}^{-1}(k)\mathbf{z}(k)$ which in this case is obtained from the summation

$$\mathbf{H}(k)^T\mathbf{R}^{-1}(k)\mathbf{z}(k) = \sum_j \mathbf{H}_j(k)^T\mathbf{R}_j^{-1}(k)\mathbf{z}_j(k). \tag{3.66}$$

Similarly,

$$\mathbf{H}(k)^T \mathbf{R}^{-1}(k) \mathbf{H}(k) = \sum_j \mathbf{H}_j(k)^T \mathbf{R}_j^{-1}(k) \mathbf{H}_j(k), \qquad (3.67)$$

for the observation information matrix. Each sensor communicates to the central processor $\mathbf{H}_j(k)^T \mathbf{R}_j^{-1}(k) \mathbf{z}_j(k)$ and $\mathbf{H}_j(k)^T \mathbf{R}_j^{-1}(k) \mathbf{H}_j(k)$. And so, the global information state vector is given by

$$\hat{\mathbf{y}}(k \mid k) = \hat{\mathbf{y}}(k \mid k-1) + \mathbf{H}(k)^T \mathbf{R}^{-1}(k) \mathbf{z}(k), \qquad (3.68)$$

and the inverse covariance (information matrix) by

$$\mathbf{P}^{-1}(k \mid k) = \mathbf{P}^{-1}(k \mid k-1) + \mathbf{H}(k)^T \mathbf{R}^{-1}(k) \mathbf{H}(k), \qquad (3.69)$$

using Equations 3.66 and 3.67. As before, the predictions $\mathbf{P}^{-1}(k \mid k-1)$ and $\hat{\mathbf{y}}(k \mid k-1)$ are given by Equations 3.55 and 3.54, respectively.

Both the above filters can be extended to non-linear estimation in the same way as was done in Section 3.3.1 for the single sensor information filter. The hierarchical filters have been presented here for illustrative purposes and they lead into the development of the decentralized information filter.

3.3.3 Decentralized Information Filter

A decentralized form of the information filter can be developed *directly* from probabilistic information update. This allows each sensor in a decentralized system to fuse observations from all the sensors in the system so as to obtain an estimate of the state of nature. Since decentralized systems are motivated by a desire for operationally robust and modular autonomous systems which are not dependent on a central processor (see page 11), the algorithm for estimating the state is expected to be equivalent to a centralized one. In other words, from an estimation optimality point of view, there are no gains in using a decentralized estimator other than the cited operational advantages.

The decentralized information filter is equivalent to the Decentralized Kalman Filter (DKF) [181][177] and indeed yields the same estimates. However, as stated earlier, in addition to its computational properties, the information filter form is desirable because it is consistent with our information update paradigm. The decentralized filter that is developed here corresponds to the Independent Likelihood Pool.

3.3 Multi-Sensor Continuous State Estimation

State Transition

The definition of a local state (Equation 3.7) is extended to that of a a state vector $\mathbf{x}_i(k)$ associated with each sensor node i. The transition of the state $\mathbf{x}_i(k)$ is described by the equation

$$\mathbf{x}_i(k) = \mathbf{F}_i(k)\mathbf{x}_i(k-1) + \mathbf{w}_i(k), \qquad (3.70)$$

where $\mathbf{F}_i(k)$ is the nodal state transition matrix and $\mathbf{w}_i(k)$ is the nodal state transition noise. The nodal state transition noise is under the usual assumptions, that is

$$\begin{aligned} E\{\mathbf{w}_i(k)\} &= 0, \\ E\{\mathbf{w}_i(k)\mathbf{w}_i(j)^T\} &= \mathbf{Q}_i(k)\delta_{k,j}, \quad \forall k,j. \end{aligned} \qquad (3.71)$$

Observation

As in Equation 3.57, the complete observation vector $\mathbf{z}(k)$ is partitioned into sub-vectors corresponding to each nodes' observation i.e. $\mathbf{z}(k) = [\mathbf{z}_1^T(k), \cdots, \mathbf{z}_N^T(k)]^T$, based on the assumption that the observation partitions are uncorrelated. The observation matrix $\mathbf{H}(k)$, and the observation noise vector $\mathbf{v}(k)$, are partitioned in a similar manner. The nodal observation equation is written

$$\mathbf{z}_i(k) = \mathbf{H}_i(k)\mathbf{x}_i(k) + \mathbf{v}_i(k). \qquad (3.72)$$

The usual assumptions are made concerning the nodal observation noise, that is, that $\mathbf{v}_i(k)$ is uncorrelated, satisfying

$$\begin{aligned} E\{\mathbf{v}_i(k)\} &= 0, \\ E\{\mathbf{v}_i(k)\mathbf{v}_i(j)^T\} &= \mathbf{R}_i(k)\delta_{k,j}, \quad \forall k,j. \end{aligned} \qquad (3.73)$$

There is also the additional assumption that the observation noise and the state transition noise are uncorrelated

$$E\{\mathbf{v}_j(k)\mathbf{w}_j(k)\} = 0, \quad \forall j \in \mathcal{N}. \qquad (3.74)$$

Information Update and Estimation

We make use of the information update for an architecture which corresponds to the Independent Likelihood Pool of Chapter 2, that is

$$l_i \stackrel{\triangle}{=} \ln p(\mathbf{x}_i \mid \{\mathbf{Z}^k\}) = \ln p(\mathbf{x}_i \mid \{\mathbf{Z}^{k-1}\}) + \sum_j \underbrace{\ln [\alpha_j \, p(\mathbf{z}_j(k) \mid \mathbf{x}_j)]}_{\text{communicated}}, \qquad (3.75)$$

where the observation set $\{\mathbf{Z}^k\}$ is as defined in Equation 2.27. From this posterior, an MMSE estimate is obtained at each sensor i as follows

$$\hat{\mathbf{x}}_i(k \mid k) = E\left\{\mathbf{x}_i(k) \mid \{\mathbf{Z}^k\}\right\}, \tag{3.76}$$

with covariance

$$\mathbf{P}_i(k \mid k) = E\left\{(\mathbf{x}_i(k) - \hat{\mathbf{x}}_i(k \mid k))(\mathbf{x}_i(k) - \hat{\mathbf{x}}_i(k \mid k))^T \mid \{\mathbf{Z}^T\}\right\}. \tag{3.77}$$

The information state vector is defined for sensor i at time-step $(j \mid l)$ as

$$\hat{\mathbf{y}}_i(j \mid l) \triangleq \mathbf{P}_i^{-1}(j \mid l)\hat{\mathbf{x}}_i(j \mid l), \tag{3.78}$$

and sensor i's information matrix is given by the inverse covariance $\mathbf{P}_i^{-1}(j \mid l)$, which is the covariance of the information state vector $\mathbf{P}_i^{-1}(j \mid l)\hat{\mathbf{x}}_i(j \mid l)$. As before, all the PDFs are assumed to be Gaussian and the derivation proceeds by following the steps outlined in Section 3.3.1, but making use of the above assumptions and l_i.

Substituting the appropriate Gaussians in l_i, evaluating $\nabla_\mathbf{x} l_i$ and using results obtained before yields

$$\begin{aligned}
&-\mathbf{P}_i^{-1}(k \mid k)(\mathbf{x}_i(k) - \hat{\mathbf{x}}_i(k \mid k)) \\
&= -\mathbf{P}_i^{-1}(k \mid k-1)(\mathbf{x}_i(k) - \hat{\mathbf{x}}_i(k \mid k)) \\
&\quad + \sum_j \mathbf{H}_j(k)^T \mathbf{R}_j^{-1}(k)(\mathbf{z}_j(k) - \mathbf{H}_j(k)\mathbf{x}_j(k)).
\end{aligned} \tag{3.79}$$

Taking $\nabla_{\mathbf{x}_i}\nabla_{\mathbf{x}_i}^T l_i$ gives the nodal Fisher information, that is, the information matrix. And so, applying $\nabla_{\mathbf{x}_i}$ again gives

$$\mathbf{P}_i^{-1}(k \mid k) = \mathbf{P}_i^{-1}(k \mid k-1) + \sum_j \mathbf{H}_j(k)^T \mathbf{R}_j^{-1}(k)\mathbf{H}_j(k). \tag{3.80}$$

Post-multiplying Equation 3.80 by $\hat{\mathbf{x}}_i(k \mid k)$ and adding to Equation 3.79 gives

$$\begin{aligned}
\mathbf{P}_i^{-1}(k \mid k)\hat{\mathbf{x}}_i(k \mid k) &= \\
\mathbf{P}_i^{-1}(k \mid k-1)\hat{\mathbf{x}}_i(k \mid k-1) &+ \sum_j \mathbf{H}_j(k)^T \mathbf{R}_j^{-1}(k)\mathbf{z}_j(k).
\end{aligned} \tag{3.81}$$

Using the definition of the nodal information state vector, Equation 3.81 can be written as

$$\hat{\mathbf{y}}_i(k \mid k) = \hat{\mathbf{y}}_i(k \mid k-1) + \sum_j \mathbf{H}_j(k)^T \mathbf{R}_j^{-1}(k)\mathbf{z}_j(k), \tag{3.82}$$

3.3 Multi-Sensor Continuous State Estimation

and the predicted covariance is given by

$$\begin{aligned}\mathbf{P}_i(k \mid k-1) &= E\{\mathbf{x}_i(k)\} \\ &\quad - E\left\{\hat{\mathbf{x}}_i(k \mid k-1)(\mathbf{x}_i(k) - \hat{\mathbf{x}}_i(k \mid k-1))^T \mid \{\mathbf{Z}^{k-1}\}\right\} \\ &= \mathbf{F}_i(k)\mathbf{P}_i(k-1 \mid k-1)\mathbf{F}_i(k)^T + \mathbf{Q}_i(k), \quad (3.83)\end{aligned}$$

based on a similar derivation as in Equation 3.39. And so, the predicted information state is given by

$$\hat{\mathbf{y}}_i(k \mid k-1) = \mathbf{P}_i^{-1}(k \mid k-1)\mathbf{F}_i(k)\mathbf{P}_i(k-1 \mid k-1)\hat{\mathbf{y}}_i(k-1 \mid k-1). \quad (3.84)$$

Partial Estimates

Partial estimates are estimates of the state based on a global prior (global information before current local observation) and current local information *only*. The importance of these partial estimates has already been alluded to, but shall become apparent in the next chapter. The partial estimate $\tilde{\mathbf{x}}_i(k \mid k)$, is obtained by considering information update given by Equation 3.11, i.e.

$$\begin{aligned}\tilde{l}_i &\triangleq \ln p(\mathbf{x}_i(k) \mid \{\mathbf{Z}^{k-1} \cup \mathbf{z}_i(k)\}) \\ &= \ln p(\mathbf{x}_i(k) \mid \{\mathbf{Z}^{k-1}\}) + \ln\left[\alpha_i\, p(\mathbf{z}_i(k) \mid \mathbf{x}_i(k))\right]. \quad (3.85)\end{aligned}$$

From this posterior an MMSE partial estimate can be obtained

$$\tilde{\mathbf{x}}_i(k \mid k) = E\left\{\mathbf{x}_i(k) \mid \{\mathbf{Z}^{k-1}\} \cup \mathbf{z}_i(k)\right\}, \quad (3.86)$$

with covariance

$$\tilde{\mathbf{P}}_i(k \mid k) = E\left\{(\mathbf{x}_i(k) - \tilde{\mathbf{x}}_i(k \mid k))(\mathbf{x}_i(k) - \tilde{\mathbf{x}}_i(k \mid k))^T \mid \{\mathbf{Z}^{k-1}\} \cup \mathbf{z}_i(k)\right\}. \quad (3.87)$$

Corresponding to this partial estimate, a *partial information state vector* at time $(j \mid l)$ can be defined, that is

$$\tilde{\mathbf{y}}_i(j \mid l) = \tilde{\mathbf{P}}_i(j \mid l)\tilde{\mathbf{x}}_i(j \mid l), \quad (3.88)$$

and a corresponding partial information matrix given by the inverse covariance $\tilde{\mathbf{P}}_i(j \mid l)$. These estimates can be obtained by following the steps outlined in Section 3.3.1, that is, evaluating (i) $\nabla_{\mathbf{x}_i}\tilde{l}_i$ and (ii) $\nabla_{\mathbf{x}_i}\nabla_{\mathbf{x}_i}^T\tilde{l}_i$. This yields the partial information state vector as

$$\tilde{\mathbf{y}}_i(k\mid k) = \hat{\mathbf{y}}_i(k\mid k-1) + \mathbf{H}_i(k)^T\mathbf{R}_i^{-1}(k)\mathbf{z}_i(k), \quad (3.89)$$

and the partial information matrix as

$$\tilde{\mathbf{P}}_i(k\mid k)^{-1} = \mathbf{P}_i^{-1}(k\mid k-1) + \mathbf{H}_i(k)^T\mathbf{R}_i^{-1}(k)\mathbf{H}_i(k), \quad (3.90)$$

where $\mathbf{P}_i^{-1}(k\mid k-1)$ and $\hat{\mathbf{y}}_i(k\mid k-1)$ are the global predictions given by Equations 3.83 and 3.84, respectively.

Decentralized Non-linear Information Filter

By considering non-linear systems a decentralized non-linear filter can also be derived. The nodal non-linear observation equation is given by

$$\mathbf{z}_i(k) = \mathbf{h}_i[k, \mathbf{x}_i(k)] + \mathbf{v}_i(k), \quad (3.91)$$

where $\mathbf{h}_i[\cdot]$ is a non-linear nodal observation model and $\mathbf{v}_i(k)$ is additive Gaussian noise. And, similarly, for the nodal state transition equation

$$\mathbf{x}_i(k) = \mathbf{f}_i[k, \mathbf{x}_i(k-1)] + \mathbf{w}_i(k), \quad (3.92)$$

where $\mathbf{f}_i[\cdot]$ is a non-linear state transition model and $\mathbf{w}_i(k)$ is additive state transition noise.

Following the, now familiar, process as before and considering only first order terms, the following filter is obtained. The information state is obtained as

$$\hat{\mathbf{y}}_i(k\mid k) = \hat{\mathbf{y}}_i(k\mid k-1) + \sum_j \nabla_{\mathbf{x}}\mathbf{h}_j[k,\hat{\mathbf{x}}_j(k\mid k-1)]^T\mathbf{R}_j^{-1}(k)\mathbf{z}'_j(k), (3.93)$$

where

$$\mathbf{z}'_j(k) = \mathbf{z}_j(k) - (\mathbf{h}_j[k,\hat{\mathbf{x}}_j(k\mid k-1)] - \nabla_{\mathbf{x}}\mathbf{h}_j[k,\hat{\mathbf{x}}_j(k\mid k-1)]\hat{\mathbf{x}}_j(k\mid k-1)). \quad (3.94)$$

The inverse covariance is given by

$$\begin{aligned}\mathbf{P}_i^{-1}(k\mid k) =\ & \mathbf{P}_i^{-1}(k\mid k-1) \\ & + \sum_j \nabla_{\mathbf{x}}\mathbf{h}_j[k,\hat{\mathbf{x}}_j(k\mid k-1)]^T\mathbf{R}_j^{-1}(k)\nabla_{\mathbf{x}}\mathbf{h}_j[k,\hat{\mathbf{x}}_j(k\mid k-1)],\end{aligned}$$

$$(3.95)$$

3.3 Multi-Sensor Continuous State Estimation

where the predicted covariance is obtained from

$$\begin{aligned}\mathbf{P}_i(k \mid k-1) &= [\nabla_\mathbf{x}\mathbf{f}_i[k,\hat{\mathbf{x}}_i(k-1 \mid k-1)]\mathbf{P}_i(k-1 \mid k-1) \\ &\times \nabla_\mathbf{x}\mathbf{f}_i[k,\hat{\mathbf{x}}_i(k-1 \mid k-1)]^T] + \mathbf{Q}_i(k),\end{aligned} \quad (3.96)$$

and the predicted information state vector is obtained from

$$\hat{\mathbf{y}}_i(k \mid k-1) = \mathbf{P}_i^{-1}(k \mid k-1)\mathbf{f}_i[k,\hat{\mathbf{x}}_i(k-1 \mid k-1)]. \quad (3.97)$$

The partial information state vector is given by

$$\tilde{\mathbf{y}}_i(k \mid k) = \hat{\mathbf{y}}_i(k \mid k-1) + \nabla_\mathbf{x}\mathbf{h}_i[k,\hat{\mathbf{x}}_i(k \mid k-1)]^T \mathbf{R}_i^{-1}(k)\mathbf{z}_i'(k), \quad (3.98)$$

and the partial inverse covariance by

$$\begin{aligned}\tilde{\mathbf{P}}_i(k \mid k) &= \mathbf{P}_i^{-1}(k \mid k-1) \\ &+ \nabla_\mathbf{x}\mathbf{h}_i[k,\hat{\mathbf{x}}_i(k \mid k-1)]^T \mathbf{R}_i^{-1}(k)\nabla_\mathbf{x}\mathbf{h}_i[k,\hat{\mathbf{x}}_i(k \mid k-1)].\end{aligned} \quad (3.99)$$

To evaluate $\nabla_\mathbf{x}\mathbf{h}_i[k,\hat{\mathbf{x}}_i(k \mid k-1)]$, the prediction $\hat{\mathbf{x}}_i(k \mid k-1)$ is obtained from

$$\hat{\mathbf{x}}_i(k \mid k-1) = \mathbf{P}_i(k \mid k-1)\hat{\mathbf{y}}_i(k \mid k-1). \quad (3.100)$$

By comparing the linear decentralized information filter and its non-linear counterpart, the following equivalences can be seen

$$\nabla_\mathbf{x}\mathbf{h}_i[\cdot] \equiv \mathbf{H}_i, \quad \text{and} \quad \nabla_\mathbf{x}\mathbf{f}_i[\cdot] \equiv \mathbf{F}_i. \quad (3.101)$$

Substituting for these equivalences makes the non-linear decentralized information filter equivalent to the linear form (with the exception of Equation 3.97). The decentralized information filter can, therefore, be summarized as shown in the box on page 80.

An information filter corresponding to the architecture described by the Independent Opinion Pool (Equation 3.8 and Equation 3.9) can also be developed in a similar manner. The resulting filter is essentially a decentralization of the equivalent hierarchy for the Independent Opinion Pool given in Section 3.2.2. However, as before, the caveat concerning the local prior information would apply to such a filter.

Decentralized Information Filter Equations

At each node i, the global information state vector is obtained as

$$\hat{\mathbf{y}}_i(k \mid k) = \hat{\mathbf{y}}_i(k \mid k-1) + \sum_j \mathbf{H}_j(k)^T \mathbf{R}_j^{-1}(k) \mathbf{z}'_j(k), \qquad (3.102)$$

where $\mathbf{z}'_j(k) = \mathbf{z}_j(k)$ for the linear filter and is given by Equation 3.94 for the non-linear filter. The information matrix is given by

$$\mathbf{P}_i^{-1}(k \mid k) = \mathbf{P}_i^{-1}(k \mid k-1) + \sum_j \mathbf{H}_j(k)^T \mathbf{R}_j^{-1}(k) \mathbf{H}_j(k). \qquad (3.103)$$

The partial information state vector is given by

$$\tilde{\mathbf{y}}_i(k \mid k) = \hat{\mathbf{y}}_i(k \mid k-1) + \mathbf{H}_i(k)^T \mathbf{R}_i^{-1}(k) \mathbf{z}'_i(k), \qquad (3.104)$$

and the partial information matrix by

$$\tilde{\mathbf{P}}_i(k \mid k)^{-1} = \mathbf{P}_i^{-1}(k \mid k-1) + \mathbf{H}_i(k)^T \mathbf{R}_i^{-1}(k) \mathbf{H}_i(k). \qquad (3.105)$$

The predictions are obtained from

$$\mathbf{P}_i(k \mid k-1) = \mathbf{F}_i(k) \mathbf{P}_i(k-1 \mid k-1) \mathbf{F}_i(k)^T + \mathbf{Q}_i(k), \qquad (3.106)$$

and

$$\hat{\mathbf{y}}_i(k \mid k-1) = \mathbf{P}_i^{-1}(k \mid k-1) \mathbf{F}_i(k) \mathbf{P}_i(k-1 \mid k-1) \hat{\mathbf{y}}_i(k-1 \mid k-1). \qquad (3.107)$$

The global estimate is obtained from

$$\hat{\mathbf{x}}_i(k \mid k) = \mathbf{P}_i(k \mid k) \hat{\mathbf{y}}_i(k \mid k), \qquad (3.108)$$

and the partial state estimate from

$$\text{and,} \quad \tilde{\mathbf{x}}_i(k \mid k) = \tilde{\mathbf{P}}_i(k \mid k) \tilde{\mathbf{y}}_i(k \mid k). \qquad (3.109)$$

3.3.4 On Initialization, Consistency and Optimality

This section has illustrated that, given an architecture in which information update is a feature of the data fusion process and given knowledge of the state transition model and the observation model,

3.3 Multi-Sensor Continuous State Estimation

an appropriate information filter can be developed directly (through the Information Update) to estimate the state.

The initialization of the information filter is a practical issue which can be approached in several ways: From considerations of maximum entropy, a least informative prior can be derived giving an initial covariance (see Appendix A.2). However, from the discussion in Chapter 2, reasonable approximations to non-informative priors can also be used. A particular advantage of the information form of the filter is that at initialization, the elements of the inverse covariance (information matrix) can be set close to *zero* implying infinite variance or highest uncertainty.

The usual concerns in estimation filters regarding consistency [16] also apply to the information filters presented here. The consistency of a filter determines the optimality of an estimate. Criteria for evaluating consistency can be summarized as follows; that the state errors and innovations are zero mean and are of the same magnitude as the corresponding covariances from the filter. This implies that the innovations should be testable as white[6]. It has already been shown by Durrant-Whyte *et al* [71] and by Willsky [208] that decentralizing or distributing the standard Kalman filter does not compromise optimality whenever the sensors are fully connected, and since the information filter is equivalent to a standard Kalman filter, the arguments for filter optimality also apply to the algorithms presented here. Non-linear filters however, somewhat compromise MMSE optimality as follows: As it has already been shown, the conditional mean achieves MMSE optimality (see Chapter 2), however, the estimate from the non-linear filter is only an approximation of the true conditional mean, i.e.

$$\hat{\mathbf{x}}(k \mid k) \approx E\left\{\mathbf{x}(k) \mid \mathbf{Z}^k\right\}.$$

And so, the corresponding $\mathbf{P}(k \mid k)$ is not an exact covariance but an approximate MSE. This arises from the use of the Taylor series expansion around the estimate for which only a finite number of terms are considered in practice.

Recent work by Grime and Durrant-Whyte [91] and Grime [90] has shown how decentralized Kalman filters can be developed for non-fully connected networks of sensor nodes.

---

[6]Techniques for testing whiteness in sequences shall be discussed in Chapter 5.

3.4 Classifying Discrete States and Objects

Discrete classification algorithms are a useful counterpart to continuous state estimation algorithms because they allow the classification of tracked objects to be inferred. This is with the *proviso* that the observed information contains information which can be used as attributes which suggest particular classifications.

3.4.1 A Bayesian Approach to Classification

In a classification or identification problem, the state vector x takes on discrete values, i.e.
$$\mathbf{x} = [X_1, X_2, \cdots, X_n], \qquad (3.110)$$
where each X_b signifies an attribute or distinct object type. Each attribute or object X_b is characterized by a set of observable parameters and so the observation model consists of a *parameter set*
$$\{M\} = \{M_1, M_2, \cdots, M_m\},$$
which relates each observable $z_b \in \mathbf{z}$ to elements of x. Using the observation model defined in Equation 2.2, the observations can be written
$$z_b = \{M\}(X_b, v_b), \qquad (3.111)$$
where v_b is the observation noise associated with each observation z_b. It is assumed that each observation z_b, of a parameter in $\{M\}$ is independent of all the other parameters in $\{M\}$. The likelihood for each possible classification $X_b \in \mathbf{x}$ is given by
$$\lambda_{\mathbf{z}(k)}(X_b) \stackrel{\triangle}{=} p(\mathbf{z}(k) \mid X_b). \qquad (3.112)$$
With the parameter set $\{M\}$, the likelihood vector for x is given by
$$\begin{aligned}\Lambda_{\mathbf{z}(k)}(\mathbf{x}) &= \left[\lambda_{\mathbf{z}(k)}(X_1), \cdots, \lambda_{\mathbf{z}(k)}(X_n)\right] \\ &= [p(\mathbf{z}(k) \mid M_1), \cdots, p(\mathbf{z}(k) \mid M_m)] \\ &\quad \times \begin{bmatrix} p(M_1 \mid X_1) & \cdots & p(M_1 \mid X_n) \\ \vdots & & \vdots \\ p(M_m \mid X_1) & \cdots & p(M_m \mid X_n) \end{bmatrix}, \end{aligned} \qquad (3.113)$$
where $p(\mathbf{z}(k) \mid \{M\})$, which can be called the *parameter model*, relates observations to the parameter set, that is, each $p(\mathbf{z}(k) \mid M_l)$ gives the

3.4 Classifying Discrete States and Objects

probability of the observation z(k) given that the true parameter to be observed is M_l. The matrix $p(\{M\} \mid \mathbf{x})$ relates the parameter model to the vector $\mathbf{x}$, that is, each $p(M_l \mid X_b)$ gives the probability that given the classification is X_b, the observed parameter is M_l. Equation 2.13 for each X_b can be written as follows

$$p(X_b \mid \mathbf{Z}^k) = p(X_b \mid \mathbf{Z}^{k-1}) \left[\alpha \, p(\mathbf{z}(k) \mid \{M\}) \, p(\{M\} \mid X_b) \right], \quad \forall X_b \in \mathbf{x}, \tag{3.114}$$

where as before, α is a normalizing constant. From this, the inferred classification is given by the *maximum a posteriori* (MAP) estimate

$$\begin{aligned} \hat{X}_b &= \arg\max_b \left[p(\mathbf{x} \mid \mathbf{Z}^k) \right] \\ &= \arg\max_b \left[p(X_1 \mid \mathbf{Z}^k), \cdots, p(X_n \mid \mathbf{Z}^k) \right]. \end{aligned} \tag{3.115}$$

From the discussion in Chapter 2, with a non-informative prior the MAP estimate is the same as the ML estimate. In general, for this algorithm an MMSE is not well defined since this is only appropriate for random parameters. However, if the posterior PDF is symmetric and the mean and mode coincide, the above MAP estimate becomes equivalent to an MMSE estimate. The algorithm of Equation 3.114 can be written in the now familiar log-likelihood form of the probabilistic information update for each $X_b \in \mathbf{x}$, i.e.

$$\begin{aligned} ln\, p(X_b \mid \mathbf{Z}^k) &= ln\, p(X_b \mid \mathbf{Z}^{k-1}) + ln\left[\frac{p(\mathbf{z}(k) \mid \{M\}) \, p(\{M\} \mid X_b)}{p(\mathbf{z}(k))} \right] \\ &= ln\, p(X_b \mid \mathbf{Z}^{k-1}) + ln\left[\alpha \lambda_{\mathbf{z}(k)}(X_b) \right], \quad \forall X_b \in \mathbf{x}. \end{aligned} \tag{3.116}$$

This classification algorithm corresponds to the single sensor architecture of Figure 3.4.

3.4.2 Hierarchical Classification

We now consider the implementation of Bayesian classification in a centralized or hierarchical manner. Such classification can be implemented on the hierarchical architectures that have been presented.

Independent Opinion Pool. In the hierarchy based on the Independent Opinion Pool and illustrated in Figure 3.5 where each sensor node communicates a complete local estimate, the log-likelihood form of the fusion equation at the central processor is given by the summation

$$ln\, p(X_b \mid \{\mathbf{Z}^k\}) = \sum_j \underbrace{ln\, p(X_b \mid \mathbf{Z}_j^k)}_{\text{communicated}}, \quad \forall X_b \in \mathbf{x}, \qquad (3.117)$$

where for each node i,

$$\begin{aligned}&ln\, p(X_b \mid \mathbf{Z}_i^k) \\ &= ln\, p(X_a \mid \mathbf{Z}_i^{k-1}) + ln\left[\frac{p(\mathbf{z}_i(k) \mid \{M_i\})\, p(\{M_i\} \mid X_a)}{p(\mathbf{z}_i(k))}\right], \quad \forall X_b \in \mathbf{x}.\end{aligned}$$
$$(3.118)$$

The central processor obtains a global classification simply by summing the local classification results from each sensor. As with the corresponding information filters, this classification algorithm is restricted to situations when local subjective information is available at each sensor, a condition rarely met in practice.

Independent Likelihood Pool. In the hierarchy based on the Independent Likelihood Pool and illustrated in Figure 3.6, the log-likelihood form of the fusion equation at the central processor is given by

$$ln\, p(X_a \mid \{\mathbf{Z}^k\}) = ln\, p(X_a \mid \{\mathbf{Z}^{k-1}\}) + \\ \sum_j \underbrace{ln\left[\frac{p(\mathbf{z}_j(k) \mid \{M_j\})\, p(\{M_j\} \mid X_{jb})}{p(\mathbf{z}_j(k))}\right]}_{\text{communicated}}, \quad \forall X_b \in \mathbf{x} \qquad (3.119)$$

In this case, the central processor updates the global classification based on the global prior and the likelihood information from the sensor observations.

3.4.3 Decentralized Classification

An equivalent decentralized algorithm can be derived which corresponds to a decentralized architecture. Such an algorithm facilitates the classification of states or attributes of states such that the inference results can be obtained at any of the sensors in the system. As a result of the decentralization, the advantages, mentioned before, of robustness, reliability and scalability follow.

3.4 Classifying Discrete States and Objects

Using the definition of a local state in Equation 3.7, the state at each sensor i is denoted

$$\mathbf{x}_i = [X_{i1}, X_{i2}, \ldots, X_{in}]. \quad (3.120)$$

And so, at sensor node i the posterior is written as $p(\mathbf{x}_i \mid \{\mathbf{Z}^k\})$ and correspondingly for the other PDFs. The posterior $p(\mathbf{x}_i \mid \{\mathbf{Z}^k\})$ is comprised of probabilities $p(X_{ib} \mid \{\mathbf{Z}^k\})$, corresponding to each object type $X_{ib} \in \mathbf{x}_i$. Each sensor node i has its own parameter model $\{M_i\}$ corresponding to its sensing device. The nodal observations are described by

$$z_{i_b} = \{M_i\}(X_{ib}, v_{i_b}). \quad (3.121)$$

Each sensor node's observation z_{i_b}, of a parameter in $\{M_i\}$ is assumed to be independent of the particular values of all the other parameters in $\{M_i\}$. The partial likelihood vector at each sensor node i, based on only its own observations, is given by

$$\Lambda_{\mathbf{z}_i(k)}(\mathbf{x}_i) = p(\mathbf{z}_i(k) \mid \{M_i\}) \ p(\{M_i\} \mid \mathbf{x}_i). \quad (3.122)$$

Each sensor node communicates its partial likelihood vectors to all connected nodes and in return receives their partial likelihoods vectors. Each sensor node is then able to fuse this likelihood information with a global prior to obtain a posterior, which is converted to a set of true probabilities by normalizing over the entire state vector $\mathbf{x}$. The fusion equation is given by

$$\begin{aligned} p(X_{ib} \mid \{\mathbf{Z}^k\}) &= p(X_{ib} \mid \{\mathbf{Z}^{k-1}\}) \\ &\times \prod_j [\alpha_j p(\mathbf{z}_j(k) \mid \{M_j\}) \ p(\{M_j\} \mid X_{jb})], \quad \forall X_{ib} \in \mathbf{x}_i. \end{aligned} \quad (3.123)$$

From which the most probable state is inferred using a MAP estimate

$$\hat{X}_i = \arg\max_b \left[p(\mathbf{x}_i \mid \{\mathbf{Z}^k\}) \right]. \quad (3.124)$$

The posterior can be written in log-likelihood form as

$$\begin{aligned} \ln p(X_{ib} \mid \{\mathbf{Z}^k\}) &= \ln p(X_{ib} \mid \{\mathbf{Z}^{k-1}\}) \\ &+ \sum_j \ln \left[\frac{p(\mathbf{z}_j(k) \mid \{M_j\}) \ p(\{M_j\} \mid X_{jb})}{p(\mathbf{z}_j(k))} \right] \\ &= \ln p(X_{ib} \mid \{\mathbf{Z}^{k-1}\}) \\ &+ \sum_j \ln \left[\alpha_j \Lambda_{\mathbf{z}_j(k)}(X_{jb}) \right], \quad \forall X_{ib} \in \mathbf{x}_i. \quad (3.125) \end{aligned}$$

As shown in Equation 3.11, each sensor node can also compute partial information. From Equation 3.123 the partial posterior at each sensor node i is given by

$$p(X_{ib} \mid \{\mathbf{Z}^{k-1}\} \cup \mathbf{z}_i(k)) =$$
$$p(X_{ib} \mid \{\mathbf{Z}^{k-1}\}) \left[\alpha_i p(\mathbf{z}_i(k) \mid \{M_i\}) \, p(\{M_i\} \mid X_{ib})\right], \quad \forall X_{ib} \in \mathbf{x}_i.$$
(3.126)

The partial MAP estimate at sensor node i denoted $\tilde{X}_i$, is given by

$$\tilde{X}_i = \arg\max_b \left[p(\mathbf{x}_i \mid \{\mathbf{Z}^k\} \cup \mathbf{z}_i(k))\right]. \qquad (3.127)$$

In terms of log-likelihoods, the partial information at each sensor node i is given by

$$ln \, p(X_{ib} \mid \{\mathbf{Z}^{k-1}\} \cup \mathbf{z}_i(k))$$
$$= ln \, p(X_{ib} \mid \{\mathbf{Z}^{k-1}\})$$
$$+ ln \left[\alpha_i p(\mathbf{z}_i(k) \mid \{M_i\}) \, p(\{M_i\} \mid X_{ib})\right], \quad \forall X_{ib} \in \mathbf{x}_i.$$
(3.128)

Only the algorithm corresponding to a decentralized architecture based on the Independent Likelihood Pool has been presented. The corresponding algorithm for the Independent Opinion Pool follows from the hierarchal algorithm defined by Equations 3.117-3.118.

3.4.4 On time varying attributes and different frames of discernment

An assumption has been made, implicitly, that the identity (or attributes) of the objects being classified does not change with time. This can be relaxed and the method generalized to allow for objects with a time-variant state classification, provided the changes are within a known set of alternatives. A transition model describing the way the true classification changes with time k, within the set of possible classifications x, is also essential. Another implicit assumption in the classification algorithm is that the target can be observed every processing cycle. This is owing to the fact that the algorithm on its own does not generate any information about the spatial positioning of the target, i.e. information which may be used to point the sensor. Thus, this algorithm can be implemented conveniently in conjunction with the information filters discussed previously.

The definition of the state as $\mathbf{x}_i$ at sensor node i allows the extension of the classification algorithms to cater for different nodal frames of discernment. In practice, this arises when sensor nodes have different sensor devices and are thus able to distinguish between different subsets of objects in the *global* frame of discernment $\mathbf{x}$. In the algorithms presented here, where the global frame is the same as the nodal frame i.e. $\mathbf{x} = \mathbf{x}_i$, $\forall i \in \mathcal{N}$, the inability of a sensor i to obtain information which permits it to classify an element $X_b \in \mathbf{x}$ is handled by assigning equi-probabilities (non-committal probabilities, see discussion in Appendix A.2) in the likelihood computations with the appropriate normalisations. However, this can become cumbersome in some applications if a particular sensor node can distinguish only a small subset of $\mathbf{x}$. The problem can be alleviated by defining a global $\mathbf{x} = \cup_i \{\mathbf{x}_i\}$, where each nodal frame of discernment $\mathbf{x}_i$, is restricted to only those targets for which the sensor obtains attribute information. In this way, computations of the local partial likelihood at i are reduced to only those of targets which can be distinguished. This makes the algorithm more modular as each sensor node (at the likelihood computation stage) need only know about targets that *it* can distinguish given its parameter model, and by implication its sensor. Such an approach is not unlike the use of reduced order states in distributed continuous estimation [6].

3.5 A Note on Data Association and Validation

While data association and validation is not addressed in detail, it is worth making a note about how this might be implemented for the algorithms which have been presented. The discussion shall be in terms of a decentralized system.

In implemented decentralized systems, it is often the case that there are several information filters or classification algorithms running at the same time. Owing to this, it becomes necessary to associate the observations with the right states (i.e. filters). In the case of the classification algorithm, a method such as the one outlined in Chapter 2 can be used satisfactorily. In the case of the information filter, the method that can be used requires some elucidation. In most Kalman filter literature, data validation gates based on considering innovation sequences are used to associate measurements and states [16]. Such a validation gate between sensor nodes i and j can be constructed as follows

$$\nu_{ij}^T(k)\mathbf{S}_{ij}^{-1}(k)\nu_{ij}(k) \leq \gamma, \qquad (3.129)$$

where γ is the gating constant and ν_{ij} is the *innovation* given by

$$\nu_{ij}(k) = \mathbf{z}_j(k) - \mathbf{H}_i(k)\hat{\mathbf{x}}_i(k \mid k-1), \qquad (3.130)$$

and the covariance of the innovation is given by

$$\mathbf{S}_{ij}(k) = E\left\{\nu_{ij}(k)\nu_{ij}^T(k)\right\} = \mathbf{R}_j(k) + \mathbf{H}_i(k)\mathbf{P}_i(k \mid k-1)\mathbf{H}_i(k)^T. \qquad (3.131)$$

Clearly, in the above there is the issue of the comparability of the innovations ν_{ij} and ν_{ji}. As suggested by Rao [178], a new validation can be constructed using a modified innovation given by

$$\eta_{ij}(k) = \mathbf{H}_j(k)^T \left[\mathbf{z}_j(k) -_j \hat{\mathbf{z}}_i(k \mid k-1)\right], \qquad (3.132)$$

such that $\eta_{ij}(k)$ can be compared to $\eta_{ji}(k)$.

The innovation can be formulated in terms of information. This can be appreciated by writing the information form of Equation 3.130 as

$$\begin{aligned}\nu'_{ij}(k) &= \mathbf{H}_i(k)^T \mathbf{R}_i^{-1}(k)\nu_{ij} \\ &= \mathbf{H}_i(k)^T \mathbf{R}_i^{-1}(k)\mathbf{z}_j(k) - \mathbf{H}_i(k)^T \mathbf{R}_i^{-1}(k)\mathbf{H}_i(k)\hat{\mathbf{x}}_i(k \mid k-1).\end{aligned} \qquad (3.133)$$

Such an approach is desirable because it makes such innovations directly compatible which the information filter algorithms. This is the approach taken by Fernandez and Durrant-Whyte [78] in their work on decentralized data validation and gating.

3.6 Bibliographical Note

Approaches other than the ones shown here have been taken for describing multi-sensor structures ranging from formal specifications given in Manyika and Durrant-Whyte [144] and the multi-sensor kernels, specifications and knowledge bases described by Henderson [105, 106] to blackboards described by Harmon [100]. Other descriptions include those by Wesson *et al* [59], Fox [83], Hu [113] and Lesser and Corkill [138].

3.6 Bibliographical Note

The literature on filtering techniques and their application to data fusion is extensive. Having ourselves drawn considerably from the texts by Bar-Shalom [16] and Maybeck [149] we recommend these for background reading on Kalman filtering. Other texts on Kalman filtering are those by Catlin [42], Scharf [188], Gelb [86] and indeed the original paper by Kalman [120]. Further examples of application of Kalman filtering and estimation techniques to data fusion problems are described in papers by Willner, Chang and Dunn [60], Willsky Bello and Castanon [208], Nandhakumar and Aggarwal [162], Alouani [6], Durrant-Whyte [69, 72] and also the edited book by Bar-Shalom [15] which has numerous examples and applications. The papers by Sandell *et al* [186], Hashemipour, Roy and Laub [104], Brown *et al* [35], Rao and Durrant-Whyte [177] Chong Mori and Chan [49] discuss distributed decentralized Kalman filtering in particular. While some these papers address optimality, the paper by Hashemipour and Laub [103] discusses precisely the issue of optimality in distributed Kalman filtering. The subject of optimality in data fusion is also addressed by Porril [173]. Other issues which also arise in distributed and decentralized systems are firstly, communication which is discussed in Speyer [197], Grime, Durrant-Whyte and Ho [91] and Manyika [142] and secondly, model distribution which is specifically discussed by Berg and Durrant-Whyte [22]. Another important consideration which has only been alluded to in this chapter is that of data association, this is discussed in considerable detail by Bar-Shalom [16], Chang, Chong and Bar-Shalom [45] where the Nearest Neighbour Standard Filter (NNSF) and the Joint Probabilistic Data Association Filter (JPDAF) are described, reference can also be made to Cox and Leonard [55].

The problems of classification and identification have been discussed much in research literature (see overview in Waltz and Llinas [203]). References to Bayesian and other probabilistic methods include Pearl [171], Berger[23], Tsitsiklis and Athans [202], Chair and Varshney [44], Rao and Durrant-Whyte [180, 179]. Another statistical approach is the Dempster Shafer method which is described by Bogler [29]. In [203], Waltz and Llinas compare the Bayesian and Dempster Shafer approaches to identification.

4 Data Fusion Management

Action will furnish belief,
but will that belief be the true one ?
This is the point you know.

- Arthur Hugh Clough (1819-1861)

Sensor management ultimately reduces to making decisions regarding alternate sensing strategies, and this chapter addresses precisely the issue of *how* to make such decisions. Sensor management decisions result in the following sequence of events; (i) decisions are implemented as sensing actions, (ii) as prescribed by these actions, sensor measurements are obtained which are expected to contain information about the state of nature, and (iii) from these measurements the state of nature is estimated or inferred, thus furnishing beliefs or knowledge about the environment. Clearly, the correctness and optimality of the result hinges on the "rationality" of the original decision-making process which prescribes the sensing actions. It is for this reason that the rationality of the decision-making process is of utmost importance.

The decision-theoretic methods presented in Chapter 2 form the basis of a *normative* approach to making sensor management decisions. A normative approach is based on an axiomatic description of the decision-making process utilising information or data relevant to the decision at hand. Such an approach can be described as being a combination of probabilistic reasoning, which establishes objectivity, and utility theory which incorporates subjectivity (and preference) into the solution. As with strict Bayesian analysis, the normative approach has traditionally been eschewed on the grounds that modelling the requisite *a priori* information, on which the decisions are based, is difficult. This is because a normative approach requires that all information used should have a "proper" probabilistic interpretation. The situation is often further exacerbated due to the fact that the solution can be hard to compute and difficult to implement practically. Notwithstanding these difficulties, a normative approach provides a general and consistent framework for formulating the management problem in decentralized systems, provided that a justifiable utility structure can be constructed which reflects the goals and preferences of the system.

We start by considering the elements that make up a normative formulation of the sensor management problem. We then propose information-based utilities and proceed to develop a decentralized solution and discuss practical issues relating to implementation.

4.1 Elements of a Normative Formulation

A normative formulation consists of; (i) the data fusion algorithms which provide information *vis-à-vis* the state(s) of nature, (ii) a framework for making management decisions, and, most importantly, (iii) a rational basis for evaluating alternate decisions which reflects the goals of the sensing system. Having already presented data fusion algorithms and a decision-theoretic framework in previous chapters, this section shall focus on the issue of the basis for evaluating decisions. The preferences of the system, which reflect its goals, are encoded in the decision structure through the utility functions. Hence, the ultimate aim is to develop such utility functions which are consistent with the aims of the sensing system. First, some of the decision theory vocabulary from Chapter 2 is stated in the context of a multi-sensor system.

4.1.1 Management Imperative, Actions and Outcomes

The basis for evaluating decisions can be thought of as the *management imperative*. In this way, the management imperative can be described as the underlying motivating purpose of the sensing system, on which any sensing activities or choices can be based and is an abstraction of more specific sensor functions and task requirements. Sensor management has as a proximate goal the optimal management of sensing resources and capabilities in order to gain information and refine knowledge about given states of nature. In a given sensor system this may be refined to include more specific requirements such as observation and track maintenance, and effective coverage of a target set. As a specific example, Popoli [172] defines the management imperative in terms of sensor functional roles and tactical benefits for a combat aircraft application.

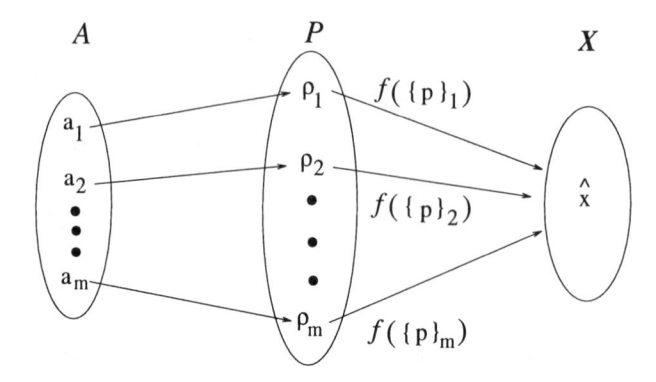

Figure 4.1: Mapping from action space through probabilistic results set to inferred state $\hat{x}$.

An *action set* is defined as a set of *distinct* sensing alternatives for achieving the management imperative. The set is denoted

$$\mathcal{A} = \{a_1, a_2, \ldots, a_m\}. \tag{4.1}$$

The actions correspond to the different sensing configurations or strategies possible. Due to uncertainty, each action a_l produces a probabilistic outcome p_l defined on $\mathcal{P}$. Mapping an action to an outcome is sometimes termed *action-outcome association*. The probabilistic outcome ρ_l is defined as the *set* of PDFs (over x) which correspond to implementing action a_l. This is written as

$$\begin{aligned} a_l \mapsto \rho_l, \quad \text{where} \quad \rho_l &\triangleq \{p_1, p_2, \ldots p_q\}_l \\ &= \{p_1, p_2, \cdots, p_q \mid a_l\}. \end{aligned} \tag{4.2}$$

Associating an action a_l with a set of distributions ρ_l over the state x, makes intuitive sense from the point of view that the sensing strategy adopted determines the probability distributions of x obtained. These distributions are subsequently used to make inferences concerning the state x, and thus the action taken has a direct bearing on the inferred state. In the inference problems considered in Chapter 3, it can be stated that a probabilistic outcome ρ_l maps through inference (estimation) to an estimate $\hat{x}$ of the state of nature as shown in Figure 4.1. The modalities of this mapping were discussed in Chapter 2, where sets of

4.1 Elements of a Normative Formulation

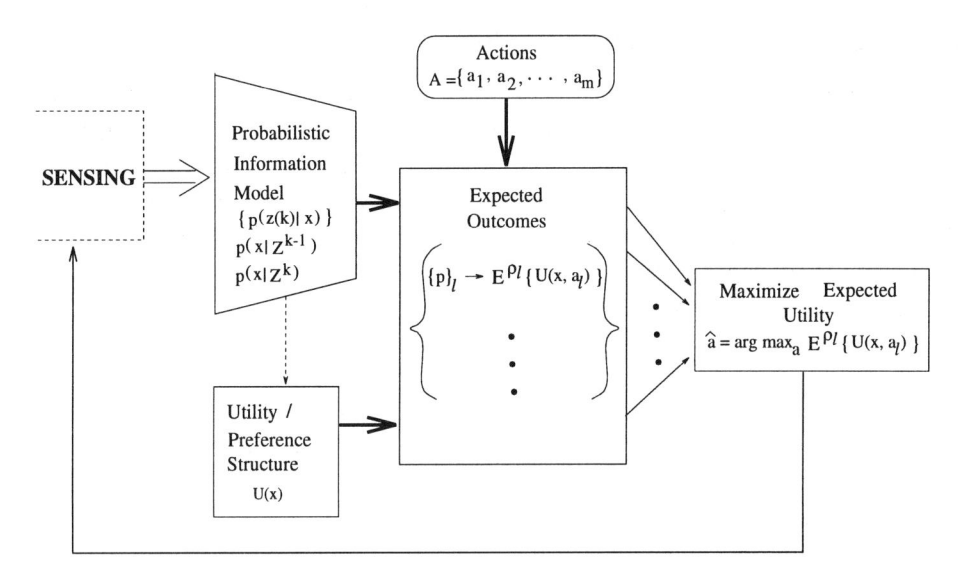

Figure 4.2: Elements of a normative approach to sensor management.

probabilities are aggregated to give a global distribution from which inferences are made.

Utility theory is used to encode the preferential structure of a sensor faced with several choices of sensing actions. A sensing system should have a preference structure which ranks the outcomes thus forming an ordered set

$$\langle \rho_1, \rho_2, \ldots, \rho_m \rangle, \quad \text{such that} \quad \rho_1 \succeq \rho_2 \succeq \cdots \succeq \rho_m, \tag{4.3}$$

where ρ_1 is the most preferred outcome. Without such a preferential structure, the management problem in terms of decision theory is ill-defined as there is no basis for evaluating actions. A utility function reflects this preferential structure through its expected value.

Decisions are then made through a maximization of the expected utilities for the various actions in $\mathcal{A}$ as shown in Chapter 2. This maximization can be written in terms of ρ_l, i.e.

$$\begin{aligned} \hat{a} &= \arg\max_a \beta(\rho_l) \\ &= \arg\max_a E^{\rho_l}\{U(x, a_l)\}. \end{aligned} \tag{4.4}$$

The above elements constitute the sensor management structure which is illustrated in Figure 4.2. Attention is now turned to the quest for appropriate utility functions.

4.1.2 The Utility of Information

The purpose of the sensor observation process, as described in Chapter 2, is the gain of information and the reduction of uncertainty. Therefore, given a choice of actions and corresponding outcomes, a rational sensor (in this sense) prefers outcomes which result in a larger gain of information. It thus seems logical to develop a utility structure based on information and its gain.

Information as Expected Utility

Information can be used as the expected utility of making sensor management decisions. For a *single* outcome PDF p_l corresponding to the action a_l, it can be written that

$$E^{p_l}\{U(\mathbf{x}, a_l)\} \propto \mathcal{I}(p_l), \tag{4.5}$$

where $\mathcal{I}(\cdot)$ is a function which quantifies the information in its argument. In general, the outcome corresponding to a_l is the *set* of distributions

$$\rho_l = \{p_1, p_2, \ldots, p_q\}_l.$$

In this case the total expected utility of action a_l is given as

$$\begin{aligned}
E^{\rho_l}\{U(\mathbf{x}, a_l)\} &= E^{p_1}\{U(\mathbf{x}, a_l)\} + E^{p_2}\{U(\mathbf{x}, a_l)\} + \cdots + E^{p_q}\{U(\mathbf{x}, a_l)\} \\
&\propto [\mathcal{I}(p_1) + \mathcal{I}(p_2) + \cdots + \mathcal{I}(p_q)] \\
&\propto \mathcal{I}(\rho_l) \\
&= \kappa \mathcal{I}(\rho_l),
\end{aligned} \tag{4.6}$$

where $\mathcal{I}(\rho_l)$ is read as the information contained in the outcome set ρ_l. The summation of expected utilities in this way is justifiable given the axioms of utility and the properties of the expectation operator. To justify the proposition in Equation 4.5, the function $\mathcal{I}(\cdot)$ must satisfy the following:

1. **Preference structure.** From utility theory, given two actions a_1 and a_2,

$$E^{p_1}\{U(\mathbf{x}, a_1)\} \leq E^{p_2}\{U(\mathbf{x}, a_2)\} \Rightarrow \rho_1 \preceq \rho_2. \tag{4.7}$$

   It is required that, given two such actions a_1 and a_2, the function $\mathcal{I}(\cdot)$ maintains same the preference structure as in Equation 4.7,

4.1 Elements of a Normative Formulation

i.e.

$$\rho_1 \preceq \rho_2 \Rightarrow \mathcal{I}(\rho_1) \leq \mathcal{I}(\rho_2). \tag{4.8}$$

And in terms of the actual resulting PDFs,

$$\rho_1 \preceq \rho_2 \Rightarrow \mathcal{I}(\{p\}_1) \leq \mathcal{I}(\{p\}_2). \tag{4.9}$$

2. **Transitivity.** Given actions a_1, a_2 and a_3, the following utility inequalities are expected

$$E^{\rho_1}\{U(\mathbf{x}, a_1)\} \leq E^{\rho_2}\{U(\mathbf{x}, a_2)\} \text{ and}$$
$$E^{\rho_2}\{U(\mathbf{x}, a_2)\} \leq E^{\rho_3}\{U(\mathbf{x}, a_3)\}, \Rightarrow \rho_1 \preceq \rho_2 \preceq \rho_3. \tag{4.10}$$

Therefore, the function $\mathcal{I}(\cdot)$ should give the same rank ordering, i.e.

$$\rho_1 \preceq \rho_2 \preceq \rho_3 \Rightarrow \mathcal{I}(\{p\}_1) \leq \mathcal{I}(\{p\}_2) \leq \mathcal{I}(\{p\}_3). \tag{4.11}$$

3. **Conditionality.** Given actions a_1, a_2 and a_3 such that in the corresponding result space $\rho_1 \preceq \rho_2$ and $\rho_2 \preceq \rho_3$, conditionality requires that

$$\alpha \, E^{\rho_1}\{U(\mathbf{x}, a_1)\} + (1 - \alpha) \, E^{\rho_3}\{U(\mathbf{x}, a_3)\}$$
$$\leq \alpha \, E^{\rho_2}\{U(\mathbf{x}, a_2)\} + (1 - \alpha) \, E^{\rho_3}\{U(\mathbf{x}, a_3)\}, \tag{4.12}$$

and so this also requires that

$$\alpha \, \mathcal{I}(\{p\}_1) + (1 - \alpha) \, \mathcal{I}(\{p\}_3) \leq \alpha \, \mathcal{I}(\{p\}_2) + (1 - \alpha) \, \mathcal{I}(\{p\}_3), \tag{4.13}$$

where $0 \leq \alpha \leq 1$.

Given these axiomatic requirements on the function $\mathcal{I}(\cdot)$, the proportionality constant in Equation 4.5 is evidently of no significance and so can be assigned $\kappa = 1$ and ignored. From the requirements, $\mathcal{I}(\cdot)$ must, therefore, be a mapping from the probability space $\mathcal{P}$ to the real line, i.e.

$$\mathcal{I}(\cdot) : \mathcal{P} \mapsto \Re, \tag{4.14}$$

which amounts to an information valuation metric. The use of such a metric is illustrated in the following example:

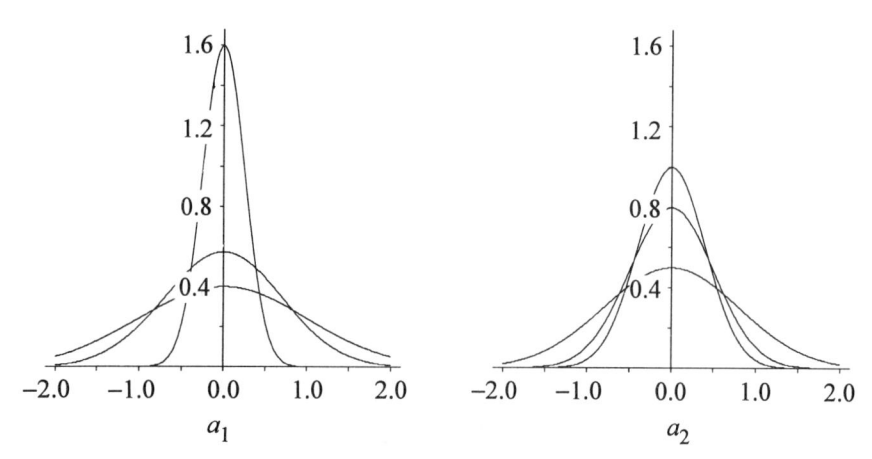

Figure 4.3: Probability distributions arising from actions a_1 and a_2 in **Example 1**.

Example 1. At any one moment, a sensor can observe 3 targets which provide information about the same state **x**. The management decision set is defined by the action set $\mathcal{A} = \{a_1, a_2\}$, where a_1 implies observation of target set $\{t\}_1$ and a_2 implies observation of target set $\{t\}_2$. The outcomes are $\rho_1 = \{p_1, p_2, p_3 \mid a_1\}$ and $\rho_2 = \{p_1, p_2, p_3 \mid a_2\}$. These PDFs are illustrated in Figure 4.3. The action resulting in higher information content as quantified by $\mathcal{I}(\rho_l)$ is chosen where

$$\mathcal{I}(\rho_l) = \mathcal{I}(p_1, a_l) + \mathcal{I}(p_2, a_l) + \mathcal{I}(p_3, a_l).$$

We propose using entropy to realize the function $\mathcal{I}(\cdot)$. Thus from Equation 4.6

$$\begin{aligned} E^{\rho_l}\{U(\mathbf{x}, a_l)\} &= \mathcal{I}(\rho_l) = [\mathcal{I}(p_1) + \mathcal{I}(p_2) \cdots + \mathcal{I}(p_q)] \\ &\triangleq [-h(p_1) - h(p_2) - \cdots - h(p_q)] \\ &= -h(\rho_l). \end{aligned} \quad (4.15)$$

The summation of entropies in this way is justified as shown in [170]. By its definition, entropy satisfies the mapping of Equation 4.14 where, for an n-dimensional discrete state vector, the space $\mathcal{P}$ is the standard simplex $(n-1)$ in $\Re^n$ [42]. And for an n-dimensional continuous state vector, $\mathcal{P}$ is the support of p such that; $supp(p) = \{x \in \Re^n \mid p(x) \neq 0\}$. Alternatively, a Fisher information metric can be used, however, entropy is preferred due to its general applicability (to discrete and continuous PDFs). A second reason for preferring entropy arises from a consideration of the resulting preference profiles.

4.1 Elements of a Normative Formulation

Preference Profiles

The behaviour of the decision-maker is characterized by the preference profile which is determined by the information-based expected utility. This behaviour can be analysed by considering Equation 4.15 as follows; for an outcome which is a single PDF p_l we have that

$$E^{p_l}\{U(\mathbf{x}, a_l)\} = E^{p_l}\{\ln p_l\}, \tag{4.16}$$

from which it follows that the utility function is given by

$$U(\mathbf{x}, a_l) = \ln p_l. \tag{4.17}$$

From the definition of strict convexity [52], it can be shown that the function $U(\mathbf{x}, a_l)$, defined as above, is convex over the distribution. This can be verified easily by showing that the second derivative is *not* non-negative everywhere. A plot of $\ln \rho$ with respect to ρ demonstrates this graphically. Ignoring the a_l's in Equation 4.17 and considering a single outcome distribution p, it can be written that for a convex (∩) function

$$\begin{aligned} \ln p(\alpha \mathbf{x}_1 + (1-\alpha)\mathbf{x}_2) &\geq \alpha \ln p(\mathbf{x}_1) + (1-\alpha)\ln p(\mathbf{x}_2) \\ U[\alpha \mathbf{x}_1 + (1-\alpha)\mathbf{x}_2] &\geq \alpha U(\mathbf{x}_1) + (1-\alpha)U(\mathbf{x}_2), \end{aligned} \tag{4.18}$$

for $0 \leq \alpha \leq 1$, which can be validated by the log-inequality. This implies that

$$U(E\{\mathbf{x}\}, a_l) \geq E\{U(\mathbf{x}, a_l)\}. \tag{4.19}$$

For the outcome which is a set of probabilities i.e. $\rho_l = \{p\}_l$, we have that (using Equation 4.6)

$$U(\mathbf{x}, a_l) = [\ln p_1 + \ln p_2 + \cdots + \ln p_q]. \tag{4.20}$$

This utility exhibits the same behaviour (i.e. Equation 4.19) as the one for the single PDF outcome.

A utility function satisfying this convex inequality results in risk-averse behaviour over the eventual inferred states **x**. This assertion is based on a proof by Ferguson [77] which demonstrates that a scalar functional can be used to order outcomes according to the decision-maker's preferences, where the precise nature of the preference structure depends on the convexity of the scalar functional. Risk-averse behaviour takes into account the probabilities associated with acquiring information, whereas risk-prone behaviour only considers the value of the information which could be gained irrespective of the probability of acquiring it. And so, this chosen utility function results in risk-aversion which befits a system where robustness and reliability are requisite characteristics.

Alternatively, the information in Equation 4.6 may be quantified using Fisher information. From the definition of Fisher information (Equation 2.61), Equation 4.5 can be written, for the case where p_l is the single distribution $p_l = p(\mathbf{Z}^k, \mathbf{x})$, as

$$E^{p_l}\{U(\mathbf{x}, a_l)\} = E\{-\nabla_\mathbf{x} \nabla_\mathbf{x}^T \ln p_l(\mathbf{Z}^k, \mathbf{x})\}. \tag{4.21}$$

To demonstrate the decision structure which results, consider the following for scalar variables

$$E^p\{U(x, a_l)\} = -E\left\{\frac{\partial^2}{\partial x^2} \ln p(z, x)\right\},$$

$$\text{thus,} \quad U(x, a_l) = -\frac{\partial^2}{\partial x^2} \ln p(z, x), \tag{4.22}$$

which is a concave function of the distribution resulting in risk-prone behaviour.

4.2 Data Fusion Information

Having established information as the expected utility, information metrics can now be developed which quantify the information in the components of the data fusion algorithms. Specifically, metrics are now presented for the information filter, the classification algorithm and their decentralized equivalents.

4.2.1 Information Filter Metrics

The logical measure of information in the information filter is Fisher information, which is simply given by the inverse covariance. In a Fisher sense, information in the posterior $p(\mathbf{x} \mid \mathbf{Z}^k)$, is given by $\mathbf{P}^{-1}(k \mid k)$ and that in the prior $p(\mathbf{x} \mid \mathbf{Z}^{k-1})$, by $\mathbf{P}^{-1}(k \mid k-1)$. The information due to the observation is given by $\mathbf{H}^T(k)\mathbf{R}^{-1}(k)\mathbf{H}(k)$. (See Chapter 3.). The discussion in Section 4.1 highlights the need for a metric which is a scalar functional of its argument and which realizes the mapping in Equation 4.14. Several techniques are available for obtaining a scalar

4.2 Data Fusion Information

functional from the Fisher information matrices[1]. However, from its definition, Fisher information is not generally applicable, being limited to cases when the distribution is continually differentiable everywhere. Given this limitation and the risk-prone decision profiles resulting from using Fisher information as expected utility, entropy is used instead as the expected utility.

In the ensuing discussion, the information in the components of the information update for the information filter and its decentralized version is quantified using entropy.

Considering a Gaussian PDF and using the result in Appendix A.1, the following is obtained:

1. **Posterior Entropy and Information.** For an n-dimensional state vector $\mathbf{x}(k)$, the entropy is given by

$$\begin{aligned} h(k) &\triangleq h(p(\mathbf{x}(k) \mid \mathbf{Z}^k)) \\ &= \frac{1}{2}E\left\{(\mathbf{x}(k) - \hat{\mathbf{x}}(k \mid k))^T \mathbf{P}^{-1}(k \mid k)(\mathbf{x}(k) - \hat{\mathbf{x}}(k \mid k)) \right. \\ &\quad \left. + \ln\left[(2\pi)^n \mid \mathbf{P}(k \mid k) \mid\right]\right\} \\ &= \frac{1}{2}\ln\left[(2\pi e)^n \mid \mathbf{P}(k \mid k) \mid\right]. \end{aligned} \qquad (4.23)$$

From this, an information metric $\mathbf{I}(k)$ is defined which quantifies all information available concerning the state $\mathbf{x}$ up to and including time-step k:

$$\mathbf{I}(k) \triangleq \mathcal{I}(p(\mathbf{x}(k) \mid \mathbf{Z}^k)) = -\frac{1}{2}\ln\left[(2\pi e)^n \mid \mathbf{P}(k \mid k) \mid\right]. \quad (4.24)$$

It must be noted that, if a non-informative prior is used at initialization, then $\mathbf{I}(k)$ quantifies all observation information up to and including that at time-step k.

2. **Prior Entropy and Information.** Similarly, the metric $\mathbf{I}(k-1)$ can be defined for information at k, given only information up to the $(k-1)$th time-step, as

$$\mathbf{I}(k-1) \triangleq \mathcal{I}(p(\mathbf{x}(k) \mid \mathbf{Z}^{k-1})) = -\frac{1}{2}\ln[(2\pi e)^n \mathbf{P}(k \mid k-1)]. \quad (4.25)$$

---

[1]The determinant | |, or a norm such as the Frobenius norm || || can be used. Work described by Jeffreys [117] on non-informative priors makes use of the determinant to quantify Fisher information.

3. **Likelihood Information.** From a consideration of the information contained in the observation, it can be shown that

$$i\left(\frac{p(\mathbf{z}(k) \mid \mathbf{x}(k))}{p(\mathbf{z}(k) \mid \mathbf{Z}^{k-1})}\right)$$
$$= -\frac{1}{2}E\left\{(\mathbf{z}(k) - \mathbf{H}(k)\mathbf{x}(k))^T \mathbf{R}^{-1}(k)(\mathbf{z}(k) - \mathbf{H}(k)\mathbf{x}(k))\right.$$
$$\left. + ln\,[(2\pi)^m \mid \mathbf{R}(k) \mid]\right\}$$
$$= -\frac{1}{2} ln\,[(2\pi e)^m \mid \mathbf{R}(k) \mid]. \qquad (4.26)$$

And so the information metric for information contained in the observation at time k *only* can be defined as

$$\mathbf{i}(k) \triangleq -\frac{1}{2} ln\,[(2\pi e)^m \mid \mathbf{R}(k) \mid]. \qquad (4.27)$$

If the observation noise matrix $\mathbf{R}(k)$ is not dependent on the actual observations, then this quantity can be pre-computed.

Similar metrics are defined for the decentralized information filter: At each sensor node i, a metric for all the *global* information up to time k is given by

$$\mathbf{I}_i(k) \triangleq \mathcal{I}(p(\mathbf{x}_i(k) \mid \{\mathbf{Z}^k\})) = -\frac{1}{2} ln\,[(2\pi e)^n \mid \mathbf{P}_i(k \mid k) \mid], \qquad (4.28)$$

and the global information metric at the kth time-step, based on only information up to the $(k-1)$th time-step, is

$$\mathbf{I}_i(k-1) \triangleq \mathcal{I}(p(\mathbf{x}_i(k) \mid \{\mathbf{Z}^{k-1}\})) = -\frac{1}{2} ln\,[(2\pi e)^n \mid \mathbf{P}_i(k \mid k-1) \mid]. \qquad (4.29)$$

A partial (local) information metric for information up to time k can also be defined as

$$\tilde{\mathbf{I}}_i(k) \triangleq \mathcal{I}(p(\mathbf{x}_i(k) \mid \{\mathbf{Z}^{k-1}\} \cup \mathbf{z}_i(k))) = -\frac{1}{2} ln\,\left[(2\pi e)^n \mid \tilde{\mathbf{P}}_i(k \mid k) \mid\right], (4.30)$$

and a metric for sensor i's observation information at time k defined as

$$\mathbf{i}_i(k) \triangleq \mathcal{I}(\alpha_i p(\mathbf{z}_i(k) \mid \mathbf{x}_i)) = -\frac{1}{2} ln\,[(2\pi e)^m \mid \mathbf{R}_i(k) \mid]. \qquad (4.31)$$

In the same way as discussed here, it is also possible to define metrics for the information corresponding to the predictions of the information filter.

4.2.2 Metrics for Bayesian Classification

Fisher information cannot be used to quantify information in the classification algorithm. This is because when x is discrete, the log-likelihood is not continually differentiable in x which is a necessary condition in the definition of Fisher information. On account of this, we use entropy as defined for discrete states in Equation 2.55.

1. **Posterior Entropy and Information.** The posterior entropy is given by

$$h(k) \triangleq E\{-\ln p(\mathbf{x} \mid \mathbf{Z}^k)\} = -\sum_b p(X_b \mid \mathbf{Z}^k) \ln \left[p(X_b \mid \mathbf{Z}^k)\right], \quad (4.32)$$

and so, the metric for all the information concerning the state x available up to and including k is given by

$$\mathbf{I}(k) = -h(k). \quad (4.33)$$

As with the information filter, it must be noted that if a non-informative prior is used at initialization, then $\mathbf{I}(k)$ represents *all* the observation information up to and including time-step k.

2. **Prior Entropy and Information.** Similarly, the prior entropy is given by

$$h(k-1) \triangleq E\{-\ln p(\mathbf{x} \mid \mathbf{Z}^{k-1})\} = -\sum_b p(X_b \mid \mathbf{Z}^{k-1}) \ln \left[p(X_b \mid \mathbf{Z}^{k-1})\right], \quad (4.34)$$

and the corresponding information metric is $\mathbf{I}(k-1) = -h(k-1)$.

3. **Likelihood Information.** The information metric for the observation information at time k is defined by

$$\mathbf{i}(k) = E\left\{\ln \left[\frac{p(\mathbf{z}(k) \mid \mathbf{x})}{p(\mathbf{z}(k))}\right]\right\} = \sum_b \frac{p(\mathbf{z}(k) \mid X_b)}{p(\mathbf{z}(k))} \ln \left[\frac{p(\mathbf{z}(k) \mid X_b)}{p(\mathbf{z}(k))}\right]$$

$$= \sum_b \alpha \Lambda_{\mathbf{z}(k)}(X_b) \ln \left[\alpha \Lambda_{\mathbf{z}(k)}(X_b)\right]. \quad (4.35)$$

In the same manner, metrics can be defined for the decentralized classification algorithm. The metric for global information at sensor node i up to time k is given as

$$\mathbf{I}_i(k) = \sum_b p(X_{ib} \mid \{\mathbf{Z}^k\}) \ln \left[p(X_{ib} \mid \{\mathbf{Z}^k\})\right], \quad (4.36)$$

and for the partial classification information at time k by

$$\tilde{\mathbf{I}}_i(k) = \sum_b p(X_{ib} \mid \{\mathbf{Z}^{k-1}\} \cup \mathbf{z}_i(k)) \, ln \, \left[p(X_{ib} \mid \{\mathbf{Z}^{k-1}\} \cup \mathbf{z}_i(k)) \right], \quad (4.37)$$

and finally, for the observed information at time k;

$$\mathbf{i}_i(k) = \sum_b \alpha_i \Lambda_{\mathbf{z}_i(k)}(X_{ib}) \, ln \, \left[\alpha_i \Lambda_{\mathbf{z}_i(k)}(X_{ib}) \right]. \quad (4.38)$$

4.3 Towards a Decentralized Sensor Management Solution

In order to avoid repetition in presenting sensor management here, the following economies are made: Firstly, the methods are posited in the context of a decentralized system on the understanding that methods for the less complex hierarchical system follow. Secondly, presentations are made in terms of results and components of the information filter, on the understanding that the same can be done for the classification algorithm unless otherwise stated. And as a corollary, the work presented is in terms of the decentralized architecture and algorithms which result from considering the Independent Likelihood Pool (see Chapters 2 and 3).

In setting out a method for decentralized sensor management, the following assumption is made:

Assumption 2 : *All sensors make conditionally independent observations in order to make inferences about the same state* **x**.

This is a simplification which allows direct comparisons of the expected utilities by ensuring that all the distributions pertain to the same state x, thus pointing clearly to the optimal action. Implicit is the assumption that all data is correctly associated.

4.3.1 Information Available to a Sensor Node

Subsequent to data fusion communication and assimilation, each sensor node i has the following probabilistic information:

1. Information directly concerning sensor i, that is

$$\mathbf{I}_i(k), \quad \mathbf{I}_i(k-1), \quad \tilde{\mathbf{I}}_i(k) \text{ and } \mathbf{i}_i(k), \quad (4.39)$$

4.3 Towards a Decentralized Sensor Management Solution

based on the global posterior and prior, the partial posterior and the likelihood respectively.

2. As a direct consequence of data fusion communication, each sensor has likelihood information from *all* the sensors in the system. This information is in the form of the log-likelihoods

$$ln\ [\alpha_j\ p(\mathbf{z}_j(k)\mid \mathbf{x}_j(k))],\quad \forall j \in \mathcal{N}$$

and in terms of the algorithms this information is represented by

$$\mathbf{H}_j^T(k)\mathbf{R}_j^{-1}(k)\mathbf{H}_j(k)\ \ \text{and}\ \ ln\ \left[\alpha_j \lambda_{\mathbf{z}_j(k)}(X_{jb})\right],\quad \forall j \in \mathcal{N} \quad (4.40)$$

for the decentralized information filter and classification algorithms respectively. Therefore, each sensor i can compute all the other sensors' observation information metrics i.e.

$$\mathbf{i}_j(k),\quad \forall j \in \mathcal{N}. \quad (4.41)$$

In addition, each sensor i is able to reconstruct the partial posteriors for all the nodes $j \in \mathcal{N}$. This is made possible by the assertion that for a fully connected system, the global posteriors and priors are identical at every node. The reconstruction is as follows; each node j computes its partial posterior according to

$$ln\ p(\mathbf{x}_j \mid \{\mathbf{Z}^{k-1}\} \cup \mathbf{z}_j(k)) = ln\ p(\mathbf{x}_j \mid \{\mathbf{Z}^{k-1}\}) + ln\ [\alpha_j\ p(\mathbf{z}_j(k) \mid \mathbf{x}_j)]. \quad (4.42)$$

The last term on the right hand side is the likelihood which is communicated by j to all the other sensors for data fusion purposes. The first term on the right hand side is the global prior which is the same at every node. Hence, sensor i after communicating with j (i.e., having received j's likelihood), has all the information necessary to compute j's partial posterior as in Equation 4.42. This is illustrated for the information filter as follows; from Equation 3.90, sensor j computes its partial information matrix according to

$$\tilde{\mathbf{P}}_j^{-1}(k \mid k) = \mathbf{P}_j^{-1}(k \mid k-1) + \mathbf{H}_j^T(k)\mathbf{R}_j^{-1}(k)\mathbf{H}_j(k).$$

Sensor node i can reconstruct j's partial information matrix as follows

$$(\text{at } i):\quad \tilde{\mathbf{P}}_j^{-1}(k \mid k) = \mathbf{P}_j^{-1}(k \mid k-1) + \underbrace{\mathbf{H}_j^T(k)\mathbf{R}_j^{-1}(k)\mathbf{H}_j(k)}_{\text{communicated}}.$$

The term unavailable at sensor node i in this equation is $\mathbf{P}_j^{-1}(k \mid k-1)$, but for a fully connected system, $\mathbf{P}_j^{-1}(k \mid k-1) = \mathbf{P}_i^{-1}(k \mid k-1)$. Thus, sensor i can reconstruct sensor j's partial information, i.e.

$$(\text{at } i): \quad \tilde{\mathbf{P}}_j^{-1}(k \mid k) = \underbrace{\mathbf{P}_i^{-1}(k \mid k-1)}_{\text{own prior}} + \underbrace{\mathbf{H}_j^T(k)\mathbf{R}_j^{-1}(k)\mathbf{H}_j(k)}_{\text{communicated}}. \quad (4.43)$$

Therefore, in addition to quantifying likelihood information from every sensor j, sensor i can, if required, quantify the partial posterior information for every sensor j. The same result can also be shown for the classification algorithm. In this way, each sensor i is able to obtain the partial information metric

$$\tilde{\mathbf{I}}_j(k), \quad \forall j \in \mathcal{N}. \quad (4.44)$$

4.3.2 The Probabilistic Implications of Actions

For a sensor system, each action $a_l \in \mathcal{A}$ gives rise to a particular sensing configuration. A particular sensing configuration or strategy determines the probabilistic information that is obtained by the sensing system. Hence each action can be said to have probabilistic implications, elsewhere, this has been termed action-outcome association. To illustrate and develop this, we present a representative example. The example also introduces and develops several concepts.

Example 2. Consider a system of 3 decentralized sensors, each of which is able to observe a single entity that we shall call a "target", with a view to estimating a state $\mathbf{x}(k)$, which in this case is the location of the platform on which the sensors are mounted. It is assumed that there are 3 targets currently being observed, that is, the target set is $\mathcal{T} = \{t_1, t_2, t_3\}$ as illustrated in Figure 4.4.

Consider the sensing scenario, as depicted in Figure 4.4, in which each sensor can observe only a single target. Thus, all possible actions are constrained by a 1-1 sensor-target assignment. In this case, the action set is defined by the various possible sensor-target assignments. These can, for example, be

$$\mathcal{A} = \begin{cases} a_1 = (t_1 \longrightarrow j;\ t_2 \longrightarrow i;\ t_3 \longrightarrow k), \\ a_2 = (t_1 \longrightarrow i;\ t_2 \longrightarrow k;\ t_3 \longrightarrow j), \\ a_3 = (t_1 \longrightarrow k;\ t_2 \longrightarrow i;\ t_3 \longrightarrow j), \\ a_4 = (t_1 \longrightarrow j;\ t_2 \longrightarrow k;\ t_3 \longrightarrow i), \\ \vdots \end{cases}. \quad (4.45)$$

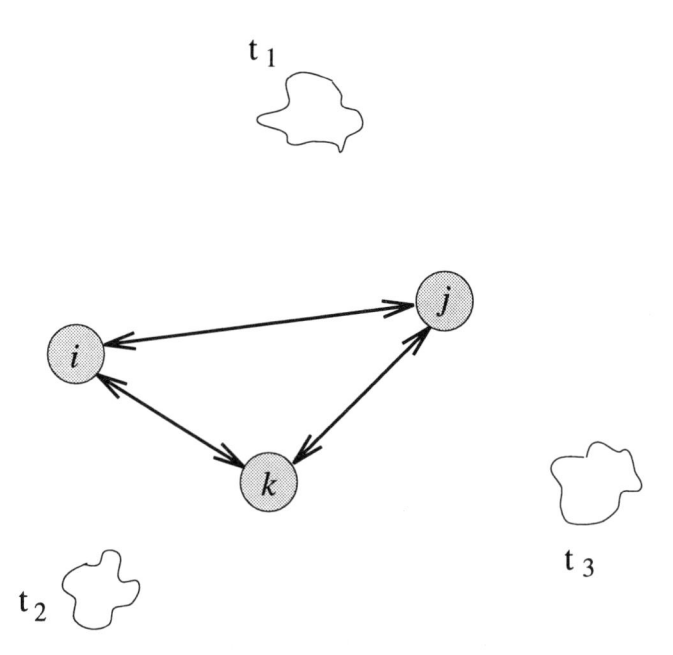

4.3 Towards a Decentralized Sensor Management Solution

Figure 4.4: The decentralized sensor system of **Example 2**. The system comprises 3 fully connected sensor nodes, i, j and k in an environment where there are 3 targets (features) to be observed.

Some of these sensor-target assignments are illustrated in Figure 4.5.

In maximizing the expected utilities in order to obtain the optimal action, we consider posterior information or equivalently likelihood information (by the Likelihood Principle, see Equations 2.41 and 2.42 and discussion in Section 2.3.1). But firstly, the notation is extended to the decentralized case. For each sensor i, the set of outcome PDFs resulting from action a_l is written ρ_{li} and given by

$$\rho_{li} \triangleq \{p_1, p_2, \ldots, p_q \mid a_l\}_i. \tag{4.46}$$

The set of all outcome PDFs for *all* the sensors arising from action a_l is written

$$\{\rho\}_l \triangleq \{\rho_{li}\} = \{\rho_{l1}, \rho_{l2}, \ldots, \rho_{lN}\}, \tag{4.47}$$

where each ρ_{li} is given by Equation 4.46.

For a problem such as in Example 2, where each sensor can only observe one target at a time, each ρ_{li} is a set with only a single distribution

(a) action a_1

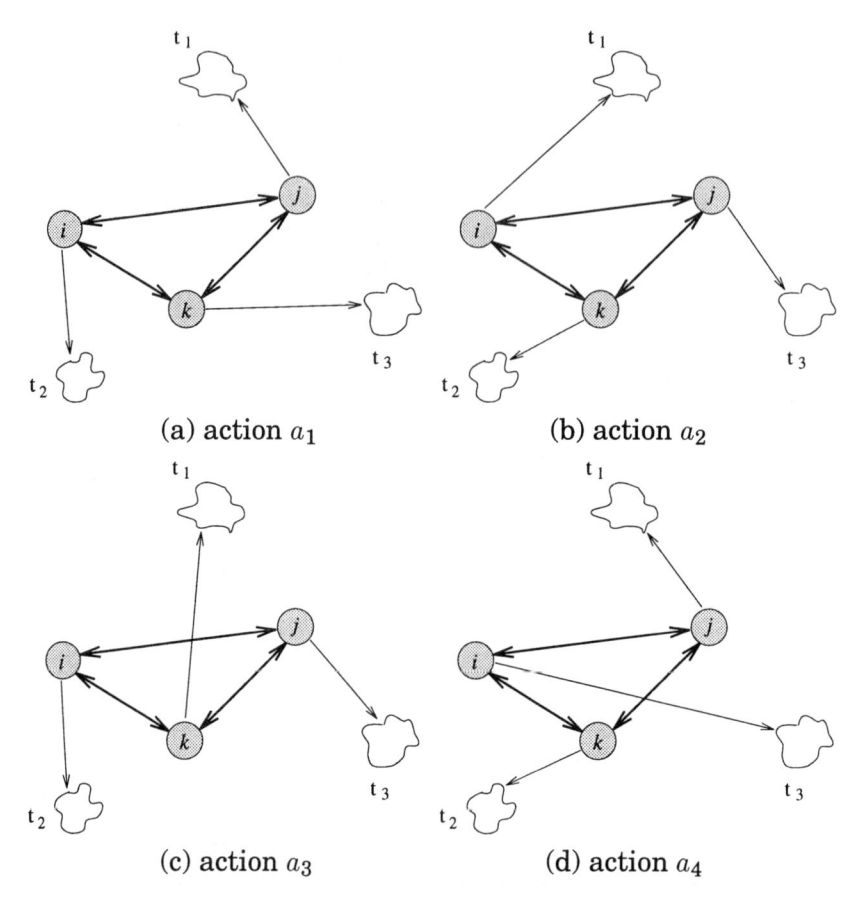

(b) action a_2

(c) action a_3

(d) action a_4

Figure 4.5: Some of the alternate sensing strategies implied by elements of the action set $\mathcal{A}$ corresponding to Figure 4.4.

p_{li}. And since management decisions can be based on either posterior information or likelihood information, using the above notation, we have that:

1. *Considering partial posterior information.* The set $\{\rho\}_l$, when the outcome at each sensor is a single PDF, is given by

$$\begin{aligned}\{\rho\}_l &= \{p_{l1}, p_{l2}, \ldots, p_{lN}\} \\ &= \left\{p(\mathbf{x}_j \mid \{\mathbf{Z}^{k-1}\} \cup \mathbf{z}_j(k)), a_l\right\}, \quad \forall j \in \mathcal{N}.\end{aligned} \quad (4.48)$$

2. *Considering likelihood information.* The set $\{\rho\}_l$, when the out-

4.3 Towards a Decentralized Sensor Management Solution

comes at each sensor are single PDFs, is written

$$\{\rho\}_l = \{p(\mathbf{z}_j(k) \mid \mathbf{x}_j), a_l\}, \quad \forall j \in \mathcal{N}. \tag{4.49}$$

Although the equivalence of these two approaches has been demonstrated in Chapter 2, there is a difference in the interpretation of the information they provide. The use of Equation 4.49 means that the only information considered as significant in the management problem is current information from observations at time-step k, that is, $\mathbf{z}_j(k)$. In contrast, Equation 4.48 considers a sensor's observation at time-step k, that is, $\mathbf{z}_j(k)$ added to all that is currently known about the state $\mathbf{x}$ at time $(k-1)$, which is based on *all* the observation information up to time $(k-1)$, that is, $\{\mathbf{Z}^{k-1}\}$. Equation 4.49 has the advantage that *no additional* computations are necessary since all the requisite information is communicated in the data fusion process. In contrast, Equation 4.48 has the disadvantage of incurring the computational overheads implicit in Equation 4.43, where the partial posteriors are re-computed at each sensor node. However, this overhead can be overcome through additional communication as shall be shown later.

Remarks

It will be noted that in Example 2, there is an implicit assumption that management starts with a system in which all targets are being observed and the question is one of allocation and re-allocation. Problems unlike the one illustrated in Example 2 also occur. Of particular interest are those which involve decisions to do with currently unobserved targets. Clearly, information such as that in Section 4.3.1 is not directly available. However, in some cases other *a priori* information may be available indirectly. In a mobile robot application, this may be information in the form of an *a priori* map describing targets (features) which are expected to be observed. In such cases, entropy measures of mutual information can be used to make allocation decisions.

More importantly, it must be noted that in the above discussion, decisions made at time step k using information metrics at k determine sensor management actions implemented at time $(k+1)$. It may be desirable to make decisions at time $(k+1)$ based on predicted information metrics at time $(k+1)$. For data fusion problems based on the information filter, it is possible to implement this by using metrics based on the filter predictions. Since the classification algorithm, as presented in this book, does not have a predictive capability, such a management strategy would not be possible. (See also discussion in Section 4.5.1.).

4.3.3 Comparable and Non-comparable Utility Solutions

The solution to the question of how to choose an optimal action from the acceptable class of actions given in Equation 2.45, depends on the admissibility of utility comparisons. The admissibility or inadmissibility of utility comparisons leads to different decisions and, consequently, a different notion of optimality. This is because the solutions adopt different maximizing philosophies.

The case for comparable information-based utilities

While conceding to Arrow's impossibility theorem as regards the general validity of utility comparisons, the following can be argued:

1. Formulation of expected utilities in terms of information quantified using entropy effectively places the expected utilities on the same relative scale. This facilitates comparison of utilities from different sensors. Furthermore, based on Assumption 2, expected utilities are in the form of information contained in the same type of PDFs concerned with the same state x for each of the algorithms. In the case of the information filter for example, the distributions are Gaussian (from Chapters 2 and 3). PDFs of the same type can be compared in terms of their informativeness with regard to the state of nature. Preference can thus be stated unambiguously, through direct comparison of expected utilities based on information metrics which satisfies the axioms in Section 4.1.2.

2. From the discussion in Chapter 2, a decentralized sensor network is unlike a human group decision problem. As seen in Equation 2.36, given that the prior at each sensor is the same (by the Likelihood Principle), differences in sensors preferences are the result of different likelihood functions (observations). This realizes the hypothetical human situation described by Savage [187], where in his justification for comparable utilities, comparisons are plausible in the case of a jury whose members "... are supposed to have common value judgements in connection with legal matters ..."[206]. Objectivity, which is implied by the likelihood function is, in this

4.3 Towards a Decentralized Sensor Management Solution

sense, a pre-requisite for direct comparison of opinions[2]. Berger [23] assures us that the likelihood function is the only basis for objectivity as it is based *only* on observed data and not subjective opinions.

3. As system designers, we find that the concept of subjective utility has a somewhat different meaning because we are able to analyse and describe it by developing sensor models. This allows us to evaluate "sensor subjectivity" quantitatively on a common scale[3]. Thus, rather than merely describing decision-makers based upon their outer exhibited preferences and behaviour, we are in the unique position of being able to "open them up" to gain an understanding of how they work. Russell and Wefald summarize this by saying "... we are interested in designing agents, rather than describing agents or being agents."[185].

More importantly, a quantitative argument further strengthening the case for comparable expected utilities, can be appreciated by considering the solutions obtained when comparable utilities are assumed and when they are not, as we now demonstrate:

From Chapter 2 and assuming comparable utilities, the optimal action is given by maximizing Equation 2.47. Here we write that the group expected utility for each action a_l is given as

$$\mathcal{B}_c(a_l) = \left\{ \sum_j w_j \left[\beta_j(a_l) - c(j) \right]^\gamma \right\}^{1/\gamma}, \qquad (4.50)$$

where $\beta_j(a_l)$ is sensor j's expected utility given by Equation 4.5. Setting $\gamma = 1$ and $w_j = 1/N$, and also ignoring individual security levels $c(j)$, the following is obtained

$$\mathcal{B}_c(a_l) = \sum_j \beta_j(a_l) = \sum_j \mathcal{I}(\rho_{lj}), \qquad (4.51)$$

---

[2]Objectivity, in this case, is simply defined as an opinion based only on observed information (data) thus avoiding the potential philosophical minefield of defining objectivity.

[3]The equivalent of this in a human situation would be to quantitatively analyse individual value judgements and how they come about, which may entail a complete psycho-analysis of an individual's whole life experiences on which judgements may be based. Objectivity in this case is impossible, given that the investigator is another such individual and not the designer!

where $\mathcal{I}(\rho_{lj})$ is read as the information contained in sensor j's set of distributions ρ_j, which result from action a_l. Assuming non-comparable utilities, the Nash solution gives the group expected utility (Equation 2.48) for each action a_l, as

$$\mathcal{B}_{nc}(a_l) = \prod_j [\beta_j(a_l) - c(j)] = \prod_j [\mathcal{I}(\rho_{lj}) - c(j)]. \quad (4.52)$$

The security level $c(j)$ is a feature of the Nash solution and is retained. While the value of $c(j)$ is arbitrary, it is usually set to be agent j's current expected utility level. Two cases arise from these considerations:

1. If decision-making occurs at the start of the observation process and a non-informative prior is used, $c(j)$ can be set to zero.

2. If sensor j is already making observations at time of decision-making, $c(j)$ is set to j's current utility level.

The different solutions that the comparable and non-comparable utility approaches give under various conditions can now be illustrated. Firstly, the example introduced in Figure 4.4 is extended.

For Example 2, using the action-outcome association suggested by Equation 4.45, values are assigned to each of the utilities for each of the sensors i, j and k. For ease of illustration, scaled values of typical expected utilities are used:

$$\begin{array}{llllllll}
 & & & & & \mathcal{B}_{c1} & \mathcal{B}_{c2} & \mathcal{B}_{nc1} & \mathcal{B}_{nc2} \\
a_1: & t_1 \xrightarrow{20} i; & t_2 \xrightarrow{16} j; & t_3 \xrightarrow{18} k; & 54 & 4 & 5760 & 0 \\
a_2: & t_1 \xrightarrow{20} i; & t_2 \xrightarrow{14} k; & t_3 \xrightarrow{15} j; & 49 & -3 & 4200 & 60 \\
a_3: & t_1 \xrightarrow{10} k; & t_2 \xrightarrow{16} j; & t_3 \xrightarrow{33} i; & 59 & 9 & 5280 & -528 \\
a_4: & t_1 \xrightarrow{10} k; & t_2 \xrightarrow{22} i; & t_3 \xrightarrow{15} j; & 47 & -3 & 3300 & 0 \\
a_5: & t_1 \xrightarrow{10} j; & t_2 \xrightarrow{22} i; & t_3 \xrightarrow{18} k; & 50 & 0 & 3960 & 0 \\
a_6: & t_1 \xrightarrow{10} j; & t_2 \xrightarrow{14} k; & t_3 \xrightarrow{33} i; & 57 & 7 & 4620 & 0
\end{array}$$

(4.53)

where $\mathcal{B}_{c1}$ and $\mathcal{B}_{nc1}$ are obtained from Equations 4.51 and 4.52 respectively. The group expected utilities $\mathcal{B}_{c1}$ and $\mathcal{B}_{c2}$, are the solutions using comparable utilities without a security level (i.e. $c(j) = 0$) and with $c(j)$ set to the current expected utility respectively. The equivalent non-comparable utility solutions are given by $\mathcal{B}_{nc1}$ and $\mathcal{B}_{nc2}$ respectively.

4.3 Towards a Decentralized Sensor Management Solution

Notice that due to the 1-1 assignment constraint, the outcomes ρ_{li} for each sensor i for a given action a_l, are single PDFs where the PDFs corresponds to the assignment of sensor i according to action a_l. The sensor target assignment at time of decision-making is assumed to be that given by action a_5.

From 4.53 the following can be observed:

1. Using the solution for comparable utilities $\mathcal{B}_{c1}$, a preferred action ordering is obtained such that;

$$\langle a_3, a_6, a_1, a_5, a_2, a_4 \rangle, \quad \text{i.e.,} \quad \hat{a}_{c1} = a_3.$$

With security levels, i.e. $\mathcal{B}_{c2}$, the following ordering is obtained

$$\langle a_3, a_6, a_1, a_5 \rangle, \quad \text{i.e.,} \quad \hat{a}_{c2} = a_3,$$

where the actions (a_2, a_4) are deemed inadmissible with respect to the sensor security levels.

2. Using the non-comparable utility solution with the security level set to $c(j) = 0$, the following preference ordering is obtained

$$\langle a_1, a_3, a_6, a_2, a_5, a_4 \rangle, \quad \text{i.e.,} \quad \hat{a}_{nc1} = a_1.$$

Increasing the security level for every sensor to the value of the least individual utility at the time of decision-making, in this case 10, gives the ordering

$$\langle a_1, a_2, (a_3, a_4, a_5, a_6) \rangle,$$

where again the optimal action is a_1, and (a_3, a_4, a_5, a_6) all result in zero group expected utilities. With security levels set to the current values of each individual sensor at the time of decision-making, the ordering is $\langle a_2, (a_1, a_4, a_5, a_6) \rangle$. The optimal action is a_2 and the actions (a_1, a_4, a_5, a_6) result in a zero group utility. In this case a_3 is an unacceptable action.

The results of Example 2 illustrate the diversity of solutions obtained by adopting either the comparable utility solution (Equation 4.51) or alternately the non-comparable utility solution (Equation 4.52).

The solutions based on comparable utilities directly maximize group expected utility without regard for individual sensor utilities. The degree of disregard depends on the security levels set, which renders some actions inadmissible. The solutions obtained by considering non-comparable utilities optimize the decision process for the individual decision-maker, which is a desirable result in a social context when making decisions regarding human individuals. However, these solutions do not result in as high group utilities as the solutions assuming comparable utilities. We can write that the total information in the system is given by

$$\mathcal{I}(\{\rho\}_{\hat{a}}) = \sum_j \mathcal{I}(\rho_{\hat{a}j}), \qquad (4.54)$$

where $\rho_{\hat{a}j}$ is the set of outcomes at sensor j, given the optimal action $\hat{a}$ as defined in Equation 4.46. By applying Equation 4.54 to the optimal solution obtained with the two approaches, it can be seen that

$$\mathcal{I}(\{\rho\}_{\hat{a}_c}) > \mathcal{I}(\{\rho\}_{\hat{a}_{nc}}),$$

where $\hat{a}_c$ and $\hat{a}_{nc}$ are the comparable and non-comparable solutions respectively. Some insight is also gained by considering the following; for the comparable utility solution, the optimal action is obtained by maximizing

$$\begin{aligned}\sum_j \beta_j(a_l) &= E^{p_1}\{U_1(\mathbf{x}, a_l)\} + E^{p_2}\{U_2(\mathbf{x}, a_l)\} + \cdots + E^{p_N}\{U_N(\mathbf{x}, a_l)\} \\ &= \int_{\mathbf{x}} U_1(\mathbf{x}, a_l) p(\mathbf{z}_1(k) \mid \mathbf{x}) d\mathbf{x} + \int_{\mathbf{x}} U_2(\mathbf{x}, a_l) p(\mathbf{z}_2(k) \mid \mathbf{x}) d\mathbf{x} + \\ &\quad \cdots + \int_{\mathbf{x}} U_N(\mathbf{x}, a_l) p(\mathbf{z}_N(k) \mid \mathbf{x}) d\mathbf{x}, \qquad (4.55)\end{aligned}$$

which from the definition of expected utility in Equation 4.6 evaluates to

$$\begin{aligned}\sum_j \beta_j(a_l) &= \mathcal{I}(p(\mathbf{z}_1(k) \mid \mathbf{x}), a_l) + \mathcal{I}(p(\mathbf{z}_2(k) \mid \mathbf{x}), a_l) + \cdots + \mathcal{I}(p(\mathbf{z}_N(k) \mid \mathbf{x}), a_l) \\ &= \sum_j \mathcal{I}(p(\mathbf{z}_j(k) \mid \mathbf{x}), a_l), \qquad (4.56)\end{aligned}$$

thus,

$$\hat{a}_c = \arg\max_{a_l} \sum_j \mathcal{I}(p(\mathbf{z}_j(k) \mid \mathbf{x}), a_l), \qquad (4.57)$$

where the assumption is made that the decisions are to be based on nodal likelihood information at time k. This can be generalized to the

4.3 Towards a Decentralized Sensor Management Solution

case when the set ρ_{lj} contains more than one element and becomes the set $\{p(\mathbf{z}_j(k) \mid \mathbf{x})\}$. In which case the action $\hat{a}_c$ is given as

$$\hat{a}_c = \arg\max_a \sum_j \beta_j(a_l) = \sum_j \mathcal{I}\left(\{p(\mathbf{z}_j(k) \mid \mathbf{x})\}, a_l\right). \quad (4.58)$$

From Equation 4.54, it is clear that $\hat{a}_c$ directly maximizes the *total* information in the system.

Conversely, adopting the solution which assumes non-comparable utilities (ignoring $c(j)$), will yield the following

$$\prod_j \beta_j(a_l) = E^{p_1}\{U_1(\mathbf{x}, a_l)\} \times E^{p_2}\{U_2(\mathbf{x}, a_l)\} \times$$
$$\cdots \times E^{p_N}\{U_N(\mathbf{x}, a_l)\}$$
$$= \int_\mathbf{x} U_1(\mathbf{x}, a_l) p(\mathbf{z}_1(k) \mid \mathbf{x}) d\mathbf{x} \times \int_\mathbf{x} U_2(\mathbf{x}, a_l) p(\mathbf{z}_2(k) \mid \mathbf{x}) d\mathbf{x} \times$$
$$\cdots \times \int_\mathbf{x} U_N(\mathbf{x}, a_l) p(\mathbf{z}_N(k) \mid \mathbf{x}) d\mathbf{x}. \quad (4.59)$$

Once again from the definition in Equation 4.6, substitutions yield

$$\hat{a}_{nc} = \arg\max_a \prod_j \beta_j(a_l)$$
$$= \arg\max_a \prod_j \mathcal{I}\left(\{p(\mathbf{z}_j(k) \mid \mathbf{x})\}, a_l\right). \quad (4.60)$$

In an information sense, the result in Equation 4.60 is somewhat meaningless and also inconsistent, since from information theory, information is additive and not multiplicative [52]. This makes questionable any approach which maximizes a product of information quantities. Consequently, the optimality of the total information resulting from maximization of a product of information terms is unclear, given total group information as the criteria for optimality.

Summary

When a sensing system is estimating the *same* state x (Assumption 2, page 102), it is evident that the solution assuming comparable utilities gives the most "optimal" solution in terms of maximizing total system information. However, in general, for a system where the state x is not the same at each node, the Nash solution is the only generally justifiable way to proceed. An example of this can be obtained by considering

Figure 4.4 and assuming that the state at each sensor node is say $\mathbf{x}_t = [x_t, y_t, z_t]$ where x_t, y_t and z_t are the spatial coordinates of the target t being observed by the sensor. Clearly, expected utilities as we have described them are not directly comparable in this case. Nevertheless, even in such a case it is possible, given good observation and sensor models, to predict and evaluate the expected utilities of sensor nodes which are not actually observing a target, and then to maximize on a summation of expected utilities for *all* targets. In this way the state can then be thought of as a single augmented one consisting of the states corresponding to each target being observed by sensor nodes in the system, thereby allowing for comparable utilities.

4.3.4 Formulation for Fusion Algorithms

For a sensor i in a decentralized system, the expected utility of tracking a target t corresponding to action a_l is given by

$$\beta_i(a_l) = \tilde{\mathbf{I}}_{it}(k), \qquad (4.61)$$

where $\tilde{\mathbf{I}}_{it}(k)$ is as given in Equation 4.30. If only observation information at time k is considered, then

$$\beta_i(a_l) = \mathbf{i}_{it}(k), \qquad (4.62)$$

where $\mathbf{i}_{it}(k)$ is as given in Equation 4.31. In general, if each sensor i is capable of observing *concurrently* several targets, then an action a_l which entails the tracking of several targets has as an outcome a set of PDFs, that is, ρ_{li}. The elements in ρ_{li} are PDFs corresponding to each target tracked by i as prescribed by action a_l. Given that sensor i observes targets in the set $\{t_1, t_2, \cdots, t_q\}$, the total expected utility at i is now given by the summation of the individual target utilities i.e.

$$\begin{aligned}
\beta_i(a_l) &= \tilde{\mathbf{I}}_{it_1}(k) + \tilde{\mathbf{I}}_{it_2}(k) + \cdots + \tilde{\mathbf{I}}_{it_q}(k) \\
&= -\frac{1}{2}ln\left[(2\pi e)^n \mid \tilde{\mathbf{P}}_{it_1}(k \mid k) \mid\right] - \frac{1}{2}ln\left[(2\pi e)^n \mid \tilde{\mathbf{P}}_{it_2}(k \mid k) \mid\right] - \\
&\quad \cdots - \frac{1}{2}ln\left[(2\pi e)^n \mid \tilde{\mathbf{P}}_{it_q}(k \mid k) \mid\right] \\
&= -\frac{1}{2}ln\left[(2\pi e)^{nq} \mid \tilde{\mathbf{P}}_{it_1}(k \mid k) \mid\; \mid \tilde{\mathbf{P}}_{it_2}(k \mid k) \mid \cdots \mid \tilde{\mathbf{P}}_{it_q}(k \mid k) \mid\right],
\end{aligned} \qquad (4.63)$$

where $\tilde{\mathbf{P}}_{it_q}(k \mid k)$ is the partial covariance when sensor i is tracking target t_q. By considering only observation information at time k, the total expected utility for sensor i is given by

$$\begin{aligned}\beta_i(a_l) &= \mathbf{i}_{it_1}(k) + \mathbf{i}_{it_2}(k) + \cdots + \mathbf{i}_{it_q}(k) \\ &= -\frac{1}{2}ln\left[(2\pi e)^{mq} \mid \mathbf{R}_{it_1}(k) \mid \; \mid \mathbf{R}_{it_2}(k) \mid \cdots \mid \mathbf{R}_{it_q}(k) \mid\right].\end{aligned} \quad (4.64)$$

Using these methods, each sensor is able to compute its own expected utility β_i for each action a_l. From the discussion in Section 4.3.3, the solution to the decentralized management problem involves the maximization of $\sum_j \beta_j(a_l)$ for each a_l. For each sensor i to compute the most optimal group action $\hat{a}$, it requires the expected utilities $\beta_j(a_l), \; \forall j \in \mathcal{N}$.

Example 2 (*cont*). In the light of the preceding discussion, each of the expected utilities for this example given in Equation 4.53 were obtained as follows: In Figure 4.4 the sensors estimate a state $\mathbf{x}(k) = [x, y]^T$ and each sensor makes observations $\mathbf{z}_{it}(k) = [z_{i1}, z_{i2}]^T$ where these observations are made with noise covariance

$$\mathbf{R}_{it}(k) = \begin{bmatrix} \sigma^2_{z_{i1}} & 0 \\ 0 & \sigma^2_{z_{i2}} \end{bmatrix}. \quad (4.65)$$

As an example,

$$\beta_i(a_1) = -const\,\frac{1}{2}ln\left[(2\pi e)^2 \mid \mathbf{R}_{it_1}(k) \mid\right] = 20.0, \quad (4.66)$$

where *const* is simply a scalar constant.

4.4 Realizing Decentralized Management

4.4.1 Computation, Communication and Bargaining

Information is required about every sensor's utilities for the various actions in order for an optimal decision to be made according to Equation 4.58. Sensor i can obtain other sensors' expected utilities either by *computing* them or having them *communicated* to it by the other sensors. However, in both cases, sensor i, in addition to evaluating its expected utility for its current tasks, has to compute its expected utility for tasks currently being performed by sensor j, $\forall j \in \mathcal{N}$. This means that if sensor j is currently observing a target t, sensor i has to evaluate its utility were it to observe target t. In order to facilitate this, information is required which gives i a basis for evaluating its utility for j's tasks. The simplest information on which such evaluations may generally be based, are the observations themselves. This can be improved upon for specific applications; for example evaluation of i's expected utility for j's tasks can be based on sector or grid information whereby i evaluates its ability to make observations in the particular sector where j is currently making observations. However, such approaches are usually related to the observations themselves and so, for purposes of this discussion, communication of the observations themselves is assumed.

The first approach considered is one in which each sensor computes every sensor's expected utility for each action. The second approach considered is one where each sensor only computes its own expected utility for each action and communicates it. These approaches are outlined assuming that management decisions are to be based only on observation information at time k (see Equation 4.49). The same methods can be used when decisions are based on observation information up to k (Equation 4.48).

Computing *all* expected utilities

In order for sensor i to compute $\beta_j(u_l)$, $\forall j \in \mathcal{N}$, the required information is $\mathbf{R}_{jt}(k)$ for each target t being tracked by each sensor j. This information is available in the form $\mathbf{H}_j(k)^T \mathbf{R}_{jt}^{-1}(k) \mathbf{H}_j(k)$ from the data fusion communications of Equation 4.40.

The maximization in Equation 4.58 over the entire action set $\mathcal{A}$ involves considering the expected utilities resulting from all the possible actions. This means that each sensor i needs to evaluate its own expected utility of tracking targets currently being observed by each sensor j, $\forall j \in \mathcal{N}$. In the simple case when each sensor i's observation noise covariance is simply $\mathbf{R}_i(k)$, these expected utilities can be pre-computed. However, in a real application the observation noise covariance may be observation dependent (and by implication also state dependent) and written as $\mathbf{R}_i[k, \mathbf{z}_i(k)]$. It is thus necessary to communicate to i the observation vector for the target for which an expected utility is to be computed. In addition, for the sensor to compute the expected utilities of the other sensors, the observation models of all the sensors must be communicated! Having computed the expected utilities of each sensor, sensor i is then able to compute the most optimal action $\hat{a}_c$. This

4.4 Realizing Decentralized Management

approach is illustrated in Figure 4.6.

The severity of the computation required with this approach can be appreciated as follows; for a system with the target set $T = \{t_1, \ldots, t_q\}$, each sensor i performs the following computations:

1. In computing its own expected utilities for the targets, sensor i evaluates the metric $\mathbf{i}_i(k)$ q times.

2. In computing the expected utilities of the other sensors for the target set T, sensor i computes the metric $\mathbf{i}_j(k)$, q times for each target and $q(N-1)$ times for all the sensors.

It is evident that there is a lot of replication in the computations performed, since each sensor repeats the same computations Nq times. In addition, the communication of observation models violates the modular and autonomous ideal of the decentralized system. The situation is exacerbated when decisions are to be based on information up to time k, that is, $\tilde{\mathbf{I}}_{jt}(k)$. This is because additional computations are required at each sensor as discussed in Section 4.3.1 to re-compute the partial posterior.

Communicating *all* expected utilities

With this approach, each sensor i computes only its *own* expected utilities $\beta_i(a_l)$, $\forall a_l \in \mathcal{A}$. In order to compute its expected utilities for targets currently being observed by other sensors the sensor i requires knowledge of their observations. Thus, the observations for all the targets are communicated by the sensor currently observing them. However, there is no need with this approach to communicate observation models. The method proceeds as shown in Figure 4.7. Each sensor computes its own expected utilities for the different actions a_l and communicates these to every other sensor in the system. On receiving the expected utilities of other nodes each sensor is thus able to compute the maximization of Equation 4.58 from which the optimal action $\hat{a}_c$ can be obtained.

The overhead incurred with this approach is the communication of expected utilities by every sensor. Since the expected utility for a given action is a numeric value, the information to be communicated by each sensor is simply a set of numbers whose cardinality depends on the number of distinct elements in $\mathcal{A}$. This, depending on system complexity, may be a smaller overhead than that incurred in the previous method

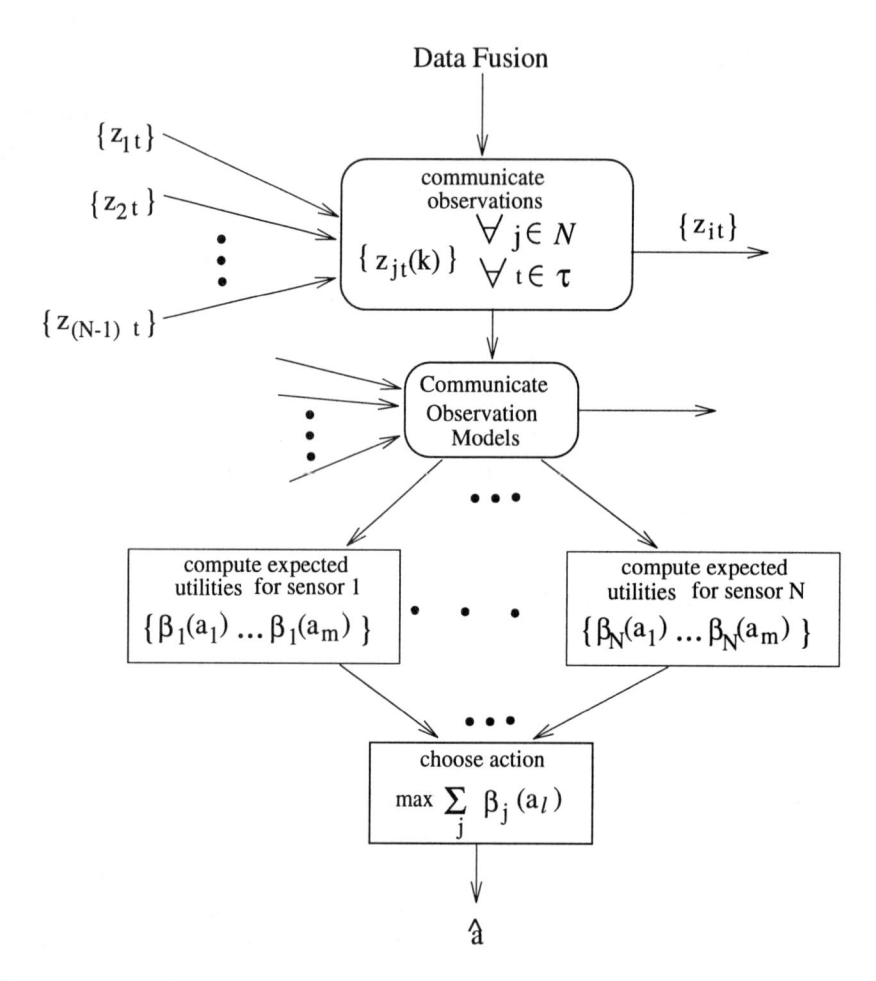

Figure 4.6: Computing the expected utilities for every sensor in the system at sensor i.

4.4 Realizing Decentralized Management

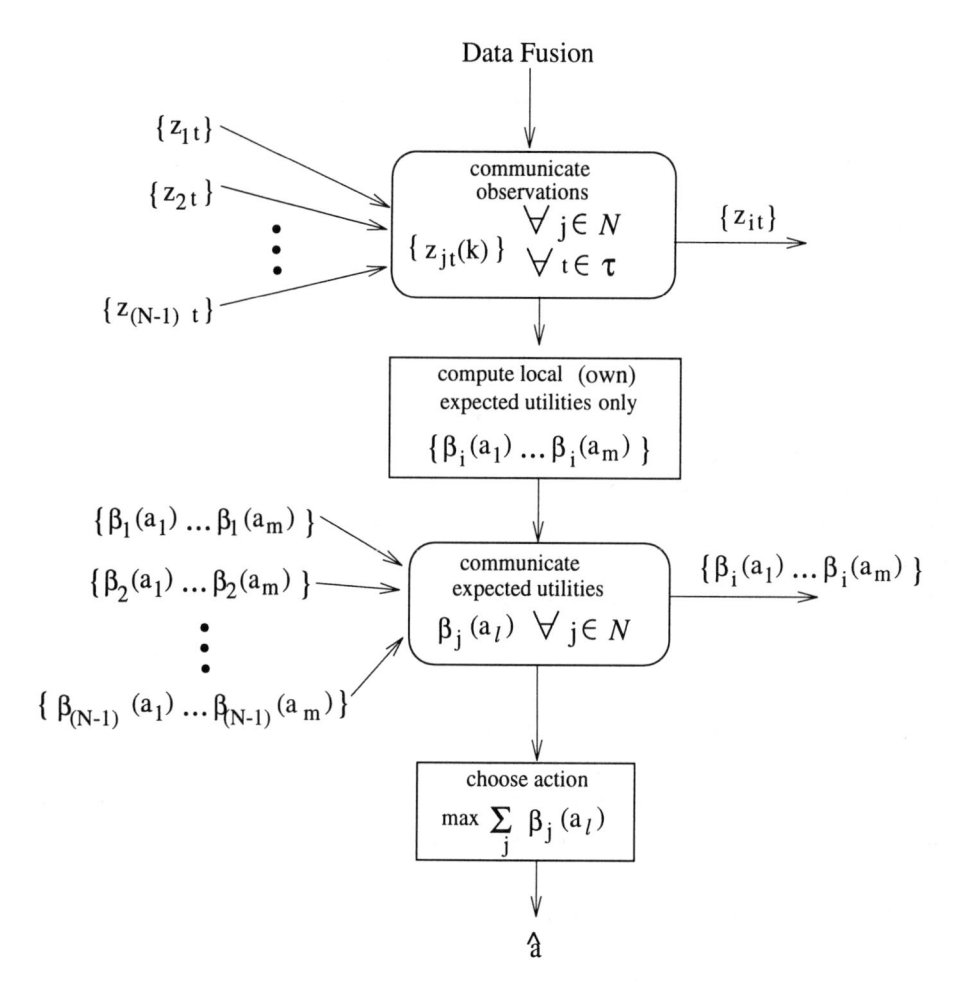

Figure 4.7: Communicating expected utilities for sensor i locally and then communicating to other sensors.

where observation models are communicated. This method also maintains the modularity of the decentralized system by keeping observation models local to each sensor. The replication in the computations of the previous method are eliminated. Whereas in Figure 4.6, for q targets, each node computes expected utilities Nq times, in Figure 4.7 each node computes this only q times.

Towards an iterative (bargaining) solution

The communication method suggests a further improvement. Suppose each sensor, by considering each $a_l \in \mathcal{A}$, computes the following ordered set i.e.

$$\left\langle \beta_i(a^{i1}), \beta_i(a^{i2}), \ldots, \beta_i(a^{ir}) \right\rangle,$$
$$\text{where } \beta_i(a^{i1}) \succeq \beta_i(a^{i2}) \succeq \cdots \succeq \beta_i(a^{ir}), \quad (4.67)$$

such that a^{i1} is sensor i's most preferred action and a^{ir} is i's least preferred action. The communication required in Figure 4.7 can be reduced if each sensor i communicates only a subset of the expected utility set corresponding to the "more" preferred actions. For example, such a subset can be chosen by considering only those actions which result in expected utilities greater than or equal to i's current expected utility i.e.

$$\{\beta_i(a_l) : \beta_i(a_l) \geq c(i)\}, \quad (4.68)$$

where $c(i)$ is sensor i's current utility at time of decision-making. Another possibility is to consider only expected utilities above a given threshold value. The effect of this is to reduce the size of the expected utility set considered and hence the number of possible actions to be considered.

In Example 2, the ordered action preferences are

$$i : \langle (a_3, a_6), (a_4, a_5), (a_1, a_2) \rangle$$
$$j : \langle (a_3, a_1), (a_4, a_2), (a_5, a_6) \rangle$$
$$k : \langle (a_5, a_1), (a_6, a_2), (a_3, a_4) \rangle,$$

where each tuple represents a pair for which the preferences, that is, the expected utilities are the same. It can be seen (Figure 4.8) that by communicating only the expected utilities corresponding to their first preference actions, the solution obtained from maximizing on these, that

4.4 Realizing Decentralized Management

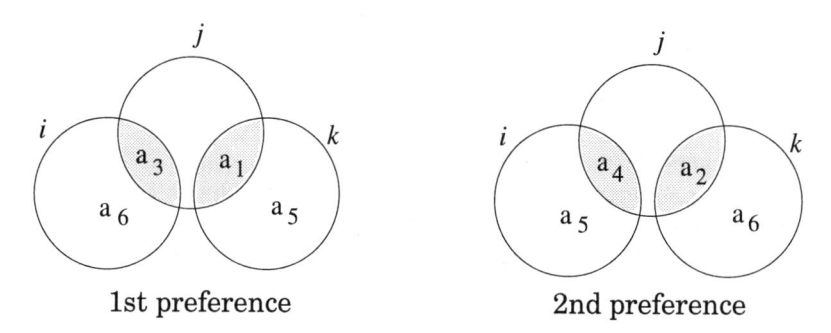

Figure 4.8: 1st and 2nd action preferences for **Example 2**.

is a_3, is the same as the one obtained by maximizing over all possible actions; a fortuitous result for this particular example. However, the idea of considering preferences in order, leads to the development of an iterative algorithm which shall now be examined.

4.4.2 An Iterative (Bargaining) Algorithm

An iterative algorithm akin to bargaining can be stated as follows:

Iterative Algorithm

1. **Order preferences.** Each sensor i computes its ordered expected utility set and, by implication, the corresponding ordered preferred action set i.e.

$$\langle \beta_i(a^{i1}), \ldots, \beta_i(a^{ir}) \rangle, \quad \Rightarrow \quad \langle a^{i1}, \ldots, a^{ir} \rangle. \tag{4.69}$$

2. **Communicate first preferences.** Each sensor i communicates its most preferred action a^{i1} and the corresponding expected utility $\beta_i(a^{i1})$ to other sensors. In the event that there are two actions with an identical expected utility (as in Example 2) then both the actions are communicated.

3. **Compare first preferences.** Each sensor i then compares its own first preference a^{i1} with other received first preferences a^{j1}, $\forall j \in \mathcal{N}$, if all refer to the same action then $\hat{a} = a^{i1}$. In practice this will be unlikely except for very simple management problems. If the communicated preferences do not refer to the same action, then each sensor i must communicate *its* expected utility corresponding to each received preferred action a^{j1}, $\forall j \in \mathcal{N}$.

4. **Maximize on first preferences.** Each sensor is then able to compute the group expected utility

$$\mathcal{B}_c(a^{j1}) = \beta_1(a^{j1}) + \beta_2(a^{j1}) + \cdots + \beta_N(a^{j1}), \quad (4.70)$$

corresponding to the most preferred action a^{j1} for each sensor j, $\forall j \in \mathcal{N}$. From this, each sensor can find the most optimal solution based on all the sensors' first preferences by maximizing over the group expected utilities corresponding to each sensors' first preference actions i.e.

$$\arg\max_{a^{j1}} \mathcal{B}_c(a^{j1}).$$

This completes the first iteration of the process.

5. **Repeat and maximize on subsequent preferences..** i.e. next communicate and compute $\mathcal{B}_c(a^{j2})$, $\forall j \in \mathcal{N}$ and so on for subsequent preferences.

The process continues until there are no more actions to consider. Iterated to completion, in other words, when all the actions in $\mathcal{A}$ have been considered, this algorithm is *exactly* equivalent to the method communicating all expected utilities. This bargaining-like[4] algorithm is illustrated in Figure 4.9.

Applying the iterative algorithm to Example 2, the first iteration yields the following; considering i's most preferred actions a^{i1}

$$\mathcal{B}_c(a^{i1}) = \mathcal{B}_c(a_3) = \beta_i(a_3) + \beta_j(a_3) + \beta_k(a_3) = 59$$
$$\mathcal{B}_c(a^{i1}) = \mathcal{B}_c(a_6) = \beta_i(a_6) + \beta_j(a_6) + \beta_k(a_6) = 57,$$

considering j's most preferred actions a^{j1}

$$\mathcal{B}_c(a^{j1}) = \mathcal{B}_c(a_3) \quad \text{(already computed)}$$
$$\mathcal{B}_c(a^{j1}) = \mathcal{B}_c(a_1) = \beta_i(a_1) + \beta_j(a_1) + \beta_k(a_1) = 54,$$

considering k's most preferred actions a^{k1}

$$\mathcal{B}_c(a^{k1}) = \mathcal{B}_c(a_1) \quad \text{(already computed)}$$
$$\mathcal{B}_c(a^{k1}) = \mathcal{B}_c(a_5) = \beta_i(a_5) + \beta_j(a_5) + \beta_k(a_5) = 50.$$

---

[4]Here the analogy with bargaining is taken to mean the process whereby each party in turn makes a bid (self-motivated) which is then considered by all and an optimum arrangement sought. The bid which maximizes the group's interests is considered the optimum one.

4.4 Realizing Decentralized Management

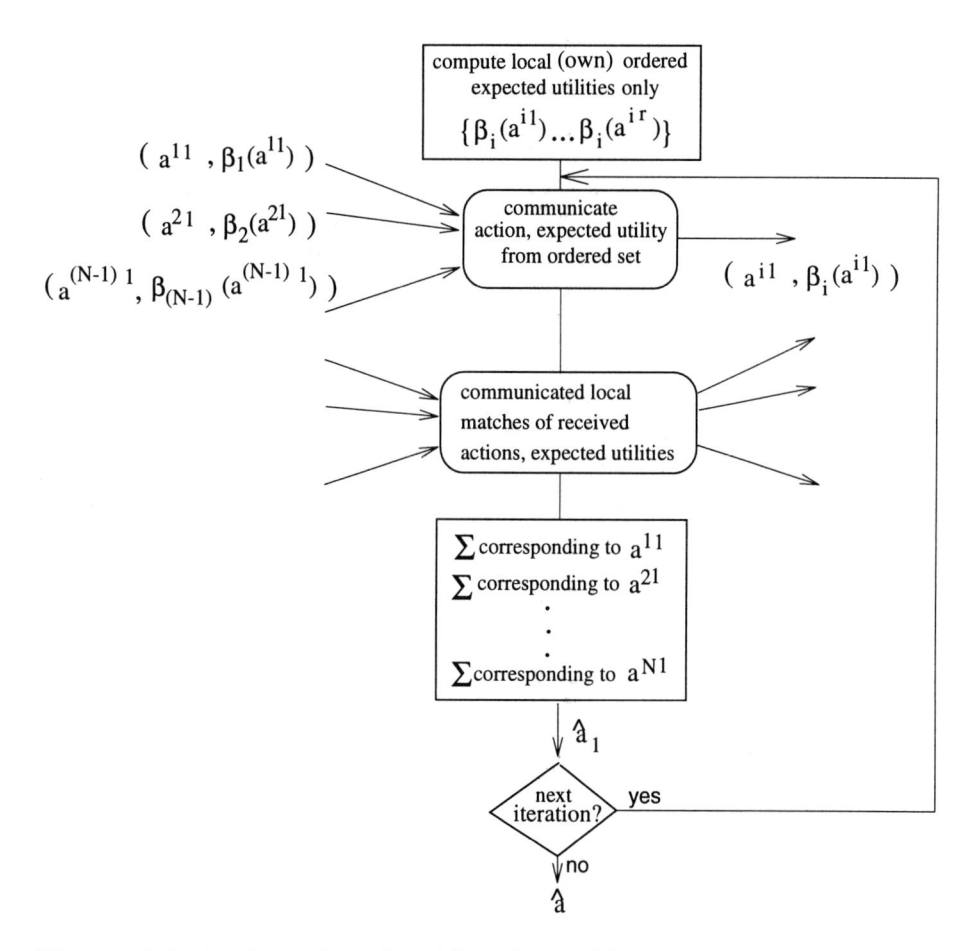

Figure 4.9: An iterative algorithm for making decentralized management decisions.

A second bargaining iteration yields

$$\mathcal{B}_c(a^{i2}) = \mathcal{B}_c(a_5) \text{ (already computed)}$$
$$\mathcal{B}_c(a^{i2}) = \mathcal{B}_c(a_4) = \beta_i(a_4) + \beta_j(a_4) + \beta_k(a_4) = 47,$$

considering j's 2nd most preferred actions a^{j2}

$$\mathcal{B}_c(a^{j2}) = \mathcal{B}_c(a_4) \text{ (already computed)}$$
$$\mathcal{B}_c(a^{j2}) = \mathcal{B}_c(a_2) = \beta_i(a_2) + \beta_j(a_2) + \beta_k(a_2) = 49,$$

considering k's 2nd most preferred actions a^{k2}

$$\mathcal{B}_c(a^{k2}) = \mathcal{B}_c(a_2) \text{ (already computed)}$$
$$\mathcal{B}_c(a^{k2}) = \mathcal{B}_c(a_6) \text{ (already computed)}.$$

Thus, a maximization based only on the first iteration in this example yields the solution $\hat{a}_c = a_3$ as before. This highlights the fact that for relatively simple management problems, there is rarely a need to have more than a single iteration of the bargaining process. However, for management problems where there are a large number of sensors and a large action set, more than one iteration may be required as the following example illustrates:

Example 3. Consider a system of 4 sensor nodes making observations of targets such that there are 8 distinct sensing configurations possible, that is, $\mathcal{A} = \{a_1, \ldots, a_8\}$. The following are the ordered sensor expected utilities for each of the actions:

sensor i	sensor j	sensor k	sensor l
$a_1 - 20$	$a_5 - 40$	$a_3 - 20$	$a_7 - 35$
$a_2 - 15$	$a_6 - 35$	$a_4 - 18$	$a_2 - 30$
$a_3 - 10$	$a_1 - 20$	$a_5 - 16$	$a_3 - 10$
$a_4 - 5$	$a_2 - 15$	$a_6 - 14$	$a_5 - 5$
$a_5 - 4$	$a_8 - 10$	$a_7 - 12$	$a_6 - 4$
$a_6 - 3$	$a_4 - 5$	$a_8 - 10$	$a_4 - 3$
$a_7 - 2$	$a_3 - 2$	$a_1 - 8$	$a_8 - 2$
$a_8 - 1$	$a_7 - 1$	$a_2 - 6$	$a_1 - 1$.

(4.71)

The 1st iteration yields

$$\mathcal{B}_c(a^{i1}) = \mathcal{B}_c(a_1) = 49$$
$$\mathcal{B}_c(a^{j1}) = \mathcal{B}_c(a_5) = 65 \longleftarrow \hat{a}$$
$$\mathcal{B}_c(a^{k1}) = \mathcal{B}_c(a_3) = 42$$
$$\mathcal{B}_c(a^{l1}) = \mathcal{B}_c(a_7) = 50.$$

4.5 A Necessary Discussion

Thus according to the first iteration the optimal action is a_5. However, proceeding to the 2nd iteration yields

$$\begin{aligned}
\mathcal{B}_c(a^{i2}) &= \mathcal{B}_c(a_2) = 66 \longleftarrow \hat{a} \\
\mathcal{B}_c(a^{j2}) &= \mathcal{B}_c(a_6) = 56 \\
\mathcal{B}_c(a^{k2}) &= \mathcal{B}_c(a_4) = 31 \\
\mathcal{B}_c(a^{l2}) &= \mathcal{B}_c(a_2) \text{ (already computed)},
\end{aligned}$$

showing that the optimal action is updated to a_2. A third iteration yields actions that have already been considered and similarly for the 4th iteration. In the 5th iteration all the actions have been considered except a_8 which yields

$$\mathcal{B}_c(a^{j5}) = \mathcal{B}_c(a_8) = 30.$$

The number of iterations required in the bargaining depends on the "degree of optimality" required in the actions taken as we now discuss.

4.5 A Necessary Discussion

4.5.1 On Rationality and Optimality

In the preceding bargaining algorithm, the question arises as to how far the bargaining or iterative process should be taken. Clearly, in the blind pursuit of "rationality" in a purely decision-theoretic sense, the iterations proceed until there are no more actions to consider. This makes the iterative algorithm equivalent to the algorithm communicating all the utilities (Figure 4.7). Given the ordered nature of the process, the chances of finding an optimal solution by considering 2nd, 3rd iterations and so on will diminish but cannot be ruled out completely. An appropriate example is that of Example 3, where the first iteration yields a group expected utility of 65 and a second iteration improves this with an action yielding 66. Two important questions arise:

- Was it worth the additional effort in computation and communication to obtain this improvement in the group expected utility? To answer this in specific terms requires knowledge of the specific system; matrix and vector sizes etc.

- How optimal is the decision-theoretic solution $\hat{a}$? The optimal action $\hat{a}$ is really only optimal at the kth time-step and, therefore, the longer it takes to compute, the less optimal it may become depending on how the expected utilities vary with time[5].

These questions arise because rationality in the decision-theoretic sense associates optimality only with the action which is finally taken rather than also taking cognizance of the costs of the actual process of arriving at the optimal solution and the possibly dynamic nature of the decision problem. Thus far we have chosen to ignore these factors and assumed that the expected utilities of Section 4.3.3 do not change in the time it takes to maximize Equation 4.58 to compute $\hat{a}$. For the ensuing discussion we depict expected utilities which are time-invariant and the more realistic time-variant expected utilities. This is illustrated in Figure 4.10 where, action $\hat{a}(k)$ is the decision-theoretic solution at time k, $\tilde{a}(k + i\xi)$ is the solution from the ith iteration of the bargaining algorithm based on probabilistic information obtained at time k and assuming that each iteration takes a fixed time ξ to compute.

For a system with a large action set, the maximization to obtain $\hat{a}$ may be computationally significant thus making ξ significant. In a dynamic system, the actual expected utilities may be significantly different at time $(k + i\xi)$ compared to what they were at time k. And indeed, as suggested by Figure 4.10, the optimal action may no longer be $\hat{a}(k)$. Under such real constraints, it thus appears reasonable to reconsider the notion of optimality and rationality. Such considerations form the basis for the concept of *Bounded Rationality* [193] which asserts that if an action takes infinite resources and time to compute then it is not a feasible action thus calling into question its optimality[6]. In the same vein, Good [89] suggests maximizing expected utilities taking into account computational and temporal costs.

And so the action obtained by truncating the iterative algorithm at a suitable point may actually be closer to the "truly" rational decision than an "optimal action" which is arrived at too late. In deciding when

---

[5]This optimality is, in practice, lost since decisions are not implemented instantaneously but at time step $(k + 1)$ using information metrics from time k. This loss of optimality can be alleviated by using predicted information metrics i.e. metrics at $(k + 1 \mid k)$.

[6]One is reminded here of the rather comic but apt example of the philosopher who on realizing that a rather large piano has been thrown out of the window of a multi-story building and is about to fall on his head, stops dead in his tracks to ponder and evaluate *all* possible actions, their outcomes and implications, in order to arrive at the rational action. A commendable approach if only time could be stopped temporarily while our thinker deliberates!

4.5 A Necessary Discussion

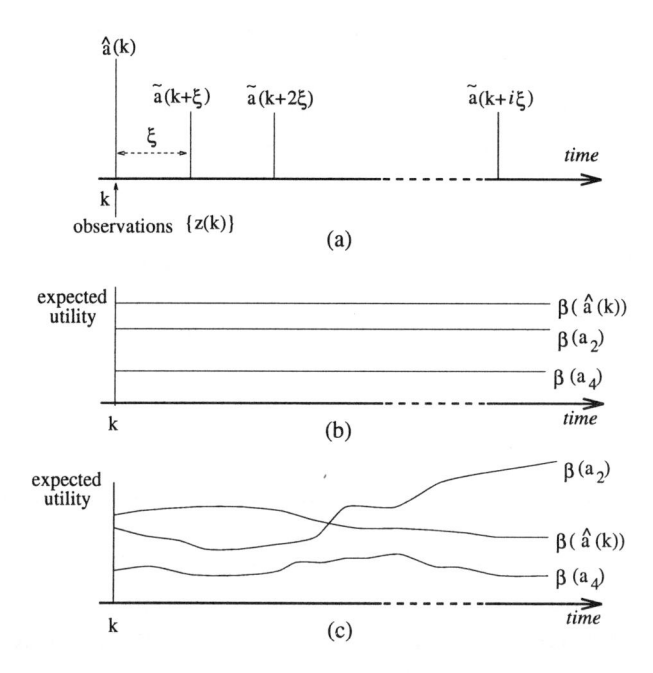

Figure 4.10: Depicting the time variations of utility functions in relation to computing more iterations of the bargaining algorithm. (a) shows the optimal actions computed at each time-step $(k+i\xi)$. (b) shows time-invariant expected utilities and (c) time variant expected utilities. From (c) it can be seen that after some time $\hat{a}(k)$, i.e. the optimal action at time k, ceases to be the optimal action.

the iterative process should terminate, several factors should be taken into account; (i) the time dependency of the expected utilities, (ii) the frequency with which management decisions are made, (iii) the computational capabilities of the decision makers, and (iv) the size of the action set and the number of sensors in the system.

4.5.2 Is it worth the bother?

Having made decisions leading to the most optimal strategy at a given time k, the action decided upon must be put into effect. An action may require considerable effort to implement depending on the application. The question arises as to whether the additional information to be gained is worth the implementation costs. This amounts to a consideration of the value of information as follows: The information to be gained while maintaining the *status quo* is compared with that to be gained by implementing the optimal action[7].

If $a_{current}$ is the sensing strategy being carried out at the time of decision-making, then the "value" of implementing the action $\hat{a}$ is given by

$$E^{\rho_{\hat{a}}}\{U(\mathbf{x},\hat{a})\} - E^{\rho_{a_{current}}}\{U(\mathbf{x},a_{current})\}. \qquad (4.72)$$

This, in fact, represents the gain in information, which can be written

$$\mathcal{I}_{gain} = \mathcal{I}(\{p(\mathbf{x} \mid \{\mathbf{Z}^k\}), \hat{a}\}) - \mathcal{I}(\{p(\mathbf{x} \mid \{\mathbf{Z}^k\}), a_{current}\}), \qquad (4.73)$$

or more generally

$$\mathcal{I}_{gain} = \mathcal{B}(\hat{a}) - \mathcal{B}(a_{current}), \qquad (4.74)$$

using the notation of Section 4.3.3. The methods of Section 4.2 can be applied to quantify this information gain, based on which a decision can be made as to whether or not to implement the optimal action. In a simple scheme, the information gain $\mathcal{I}_{gain}$ can be compared to a threshold value to determine whether action $\hat{a}$ is worth implementing.

Alternatively, distance functions such as the Hellinger distance[182], which measure "distances" between distributions, can be used to determine the difference between distributions resulting from taking action and those resulting from not taking action.

4.5.3 Coupled Management of Data Fusion Algorithms

The management methods developed are applicable to *both* the information filter and the classification algorithm even though the presentation has largely been in terms of the information filter. It is often beneficial in certain applications to implement both the information filter and the classification algorithm on the same sensor system. In such systems it is equally desirable to *couple* (or combine) the management of the two algorithms in a single management strategy. The ease with which this

---

[7]This idea of considering the value of the additional information to be gained by implementing a decision is similar to the discussion of Bounded Rationality (Section 4.5.1) when discussing the iterative algorithm. In considering the value of information, a distinction can be made between additional information, such as would be gained by enacting a management decision, and manipulation of already acquired information such as in the iterative algorithm. Clearly, the relative importance of each of these depends on the costs involved in each. For the interested reader, this is a subject discussed by Raiffa [175] and Howard [112] and more recently Russell and Wefald [185].

4.5 A Necessary Discussion

can be achieved depends largely on a consideration of Assumption 2 and is described in the following two cases.

Case 1: Whenever Assumption 2 can be guaranteed for both the classification and the information filter, the management of the two algorithms can be coupled based on developing a single information utility as follows: The information up to time-step k now becomes a linear combination of that from the information filter and that from the classification algorithm i.e.

$$\tilde{\mathbf{I}}_i(k) = \kappa_1 \, \tilde{\mathbf{I}}_i(k)_{est} + \kappa_2 \, \tilde{\mathbf{I}}_i(k)_{class}, \qquad (4.75)$$

where $\mathbf{I}_i(k)_{est}$ and $\mathbf{I}_i(k)_{class}$ is the information from the information filter and the classification respectively. The constants κ_1 and κ_2 reflect the relative importance of the information from the two algorithms. Similarly, for observation information at time k

$$\mathbf{i}_i(k) = \kappa_1 \, \mathbf{i}_i(k)_{est} + \kappa_2 \, \mathbf{i}_i(k)_{class}. \qquad (4.76)$$

To take an example, consider a system that estimates the spatial positions of targets and at the same time attempts to classify them. For this example, the information filter state may be defined as $\mathbf{x} = [x_t, y_t]^T$, where (x_t, y_t) are the 2-dimensional spatial coordinates of the target. The state for the classification algorithm may be defined as $\mathbf{x}_t = [X_1, X_2]$ where X_1 and X_2 represent target types such as $[car, human]$ for example. Because the information filter and the classification algorithm are making inferences *apropos* states of the same target t, the information components may be coupled as in Equations 4.75 and 4.76. Therefore, the information of Equations 4.75 and 4.76 can be interpreted as representing *all* the information known about target t. The role of sensor management now becomes one of optimizing the gain of both location and identity information of the target.

Case 2: Difficulties arise in attempting to couple the management of data fusion algorithms where the state of nature in the one algorithm is not related to the same entity as in the other algorithm. An illustrative example of this is a sensor making observations of a target t whose position is known, in order to estimate the sensor platform's (x_s, y_s) location in space and at the same time trying to classify attributes of the target under observation. In this case, the state of nature according to the information filter may, for example, be $\mathbf{x}_s = [x_s, y_s]^T$. The state of nature in the classification algorithm relates to the different possible attributes of t such as $\mathbf{x}_t = [plane, corner]$ for example. Thus, the information metrics $i_i(k)_{est}$ and $i_i(k)_{class}$ are not directly related and so optimizing decisions based on their linear combination is somewhat meaningless! Under such circumstances it is not clear how to proceed with a coupled solution. One approach is to manage only one algorithm, the choice of which is dependent on which type information is deemed to be of primary importance.

In the remainder of this book, the management problem shall be kept *decoupled* (in the above sense) whenever the sensing situation is as described in Case 2. The coupled approach of Equations 4.75 and 4.76 shall be used only when the sensing situation is as described in Case 1.

4.5.4 Summary

The methods presented in this chapter have shown the following:

- Sensor nodes in a decentralized system can make non-conflicting decisions *locally*, concerning sensing actions.

- The actions chosen are guaranteed to be *consistent* throughout the system and result in sensor synergy.

- The action chosen is *rational* in that it maximizes the information obtained by the system with regard to the state of nature.

The methods presented are applicable to a wide variety of systems including the specific mobile robot application that will be discussed. For this application, we now have the requisite tools to implement decentralized data fusion and manage the sensors in such a system. Thus far, we have been assuming sensors capable of offering several sensing alternatives, and that the information from such sensors is well understood and objectively modelled in terms of probabilities (a pre-requisite for employing normative techniques). The next chapter focusses precisely on these issues, developing a model for a sensor which can be managed and describing the information from it in probabilistic terms. This leads to implementation of data fusion and sensor management.

4.6 Bibliographical Note

There are numerous references to descriptive (non-normative) sensor management techniques based on which much of the implemented sys-

4.6 Bibliographical Note

tems operate. Reference, in this regard, can be made to the surveys by Waltz and Llinas [203] and Popoli [172] and that in Chapter 1. However, distributed and decentralized systems are generally *not* mentioned. It is also interesting to note that in much of the publications, the only multi-sensor systems that are managed are radar tracking systems (see [25, 26][81][99][152][207]). There is a paucity of reference material where use is made of normative methods for management in distributed and decentralized systems. The marginal use that has been made of normative methods has been in centralized and hierarchical systems. The background for much of the material presented here can be found in the literature on Decision Theory, particularly decentralized or distributed decision theory, sometimes called multi-agent decision theory. Much of this theory can be found in Berger [23], Ferguson [77], Weerahandi and Zidek [206] and Pearl [171] and Kenny and Raiffa [121]. Our use of information-based utilities is not unlike that described by Blackman [26] where use is made of Kalman filter covariance matrices. Our use of entropy based information measures as described in Manyika and Durrant-Whyte [143] is similar to the use of entropy functions by Pugh and Noble [174]. The discussion on rationality is motivated by the work of Russell and Wefald [185].

5 A Sensor Model

Give us the tools, and we will finish the job
- Winston Churchill (1874-1965)

Ultimately, multi-sensor systems rely on sensors and are thus only as good as the sensors they employ. Sensors are the tools used to make observations or measurements of physical quantities such as temperature, range, angle etc, thus providing information that may be used for inference and perception. In the parlance of the framework presented in Chapter 2, we require to express sensor information in probabilistic terms, as the likelihood function with respect to the state of nature. A sensor model entails developing an understanding of the sensed environment, the nature of the measurements provided by the sensor, the limitations of the sensor and, most importantly, probabilistic understanding of the sensor in terms of measurement uncertainty and informativeness. Moreover, the sensor model is also concerned with optimizing the information gathering activities of the sensor and reducing measurement uncertainty. All this is undertaken with the express aim of providing the most informative likelihood function, given a particular sensor, its capabilities and limitations.

This chapter is motivated, in part, by a desire to develop and demonstrate a comprehensive sensor model, hence the choice of one sensing modality for discussion. Notwithstanding the fact that the chapter is written in terms of sonar, techniques are presented which can be applied to sensor modelling for a variety of sensing modalities. Thus read as a generic sensor modelling chapter, emphasis is on:

- Gaining a physical understanding of the nature and interpretation of sensor measurements.

- Efforts at improving the operation and informativeness of a sensor.

- The measurement and analysis of uncertainty in the measurements provided by a sensor. The validation of assumptions *apropos* observation uncertainty. These assumptions are usually made in the theoretical development of data fusion and inference algorithms.

- Understanding the operation and performance constraints of a sensor, with a view to using and managing it in a multi-sensor system.

5.1 Sonar in Mobile Robotics

One application of the theoretical work in this book is in mobile robotics. A particularly active area of mobile robotics research lies in the development of autonomous robot vehicles capable of functioning in environments which may or may not be known *a priori*. Such an autonomous vehicle requires up to date information concerning the environment and the robot's relation to it. This information then allows the robot to avoid obstacles, locate itself and build models and functional descriptions of its environment. To achieve this functionality, the robot obtains the requisite information concerning the external world through sensory perception.

Current research on autonomous robot vehicles, makes use of a variety of sensor technologies and techniques, ranging from infra-red devices, laser range-finders, sonar, to CCD cameras and gyroscopes. While sonar, in the form of the standard Polaroid device [51], is the most widely used sensor in mobile robotics, its measurements do not lend themselves, as easily as would be expected, to interpretation. Although research has produced sophisticated sonar sensors capable of complete imaging, the standard Polaroid sensor, due to its extremely low cost and ease of use, remains a popular choice in mobile robotics research.

Sonar sensing has been used in mobile robotics for a wide range of tasks such as obstacle detection and avoidance [113][30], navigation and map-building [57][137]. The application of sonar to these tasks poses some difficulties. In particular, it is well known that range returns from sonar sensors are subject to distorting effects caused by long wave-lengths and wide beam-widths, making it difficult to apply any "ray-tracing" type algorithms to the interpretation of sonar data. Furthermore, because the rate at which range information can be obtained is physically limited by the speed of sound in air, a vehicle using sonar sensors to obtain map or guidance information is typically forced to adopt a "stop-look-move" motion strategy during task execution. For a survey of sonar in mobile robotics literature, reference can be made to that in Leonard and Durrant-Whyte [137].

5.2 Understanding the Sensor

5.2.1 Sonar Measurements and Interpretation

Sonar works by transmitting an acoustic signal which is reflected back on encountering an obstacle (feature) as illustrated in Figure 5.1 (a). The receiving system detects only the first echo which exceeds the threshold setting of the receiving circuit and ignores subsequent echoes. A more detailed description of the principle is given in Appendix B.1. The standard Polaroid device has a measuring range spanning from 0.30m to 10m, a nominal range accuracy of approximately 1% and a beam-width of approximately $10°$ at the -3dB level. In most indoor applications, the environment features encountered normally are planes, corners, edges and cylinders (or arcs). For features such as planes and corners, the received echo is largely due to reflection (*reflective features*). For cylinders whose radius is very small compared to the range[1], and for edges the received echo is due to diffraction (*diffractive features*).

In developing a model, we exploit the observation made by Kuc and Siegel [127] that at the wave-lengths typical of in-air sonar, common indoor environments consist predominantly of "mirror-like" reflectors or specular surfaces which correspond to features in the environment which reflect acoustic energy. Such mirror-like reflectors may be walls, cupboards, desks etc[2]. This results in range-bearing scans composed of regions in the form of circular arc segments, within which the measured

---

[1]Strictly, the comparison of cylinder radius is with the wavelength of the acoustic signal.

[2]There is a terminology issue here; "mirror-like" reflectors have also been called specularities in the literature [30][134][124]. We take the convention that, the mirror-like reflectors, result from the highly differing acoustic impedances of air and solids and, hence, the negligible sound penetration into the object due to the impedance difference [158]. Therefore, such mirror-like reflectors reflect acoustic energy in a similar manner to light reflection on mirrors. In some other literature, specularities are described as due to the wavelengths of in-air sonar, the wide beam-widths and the surface roughness in comparison with wavelength.

5.2 Understanding the Sensor

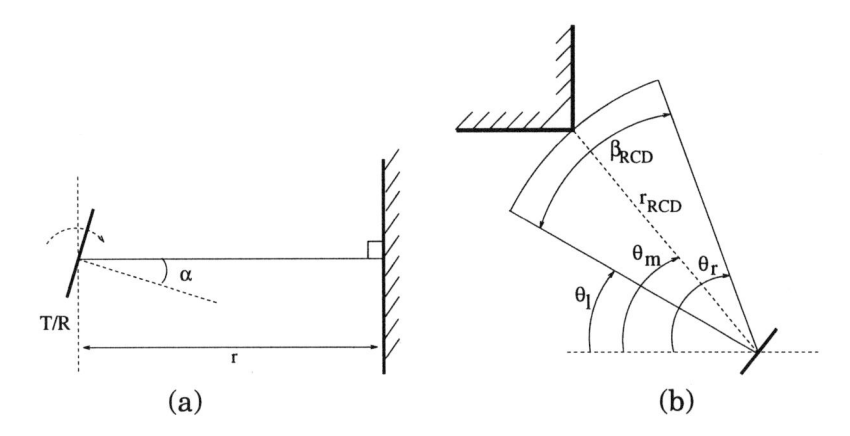

Figure 5.1: (a) Illustrates angle of inclination to the direction of propagation α and (b) parameters in the RCD model.

range is constant. These arc segments can be considered as regions in which the depth is constant. Given knowledge of these *Regions of Constant Depth* (RCD), accurate (sub-centimetre) and reliable maps of indoor environments may be constructed from sonar data [126][137].

An RCD is "generated" in the following way; when a scanning sonar device is aimed perpendicular to an environment feature, the range returned is indeed the true range to that feature. However, as the sonar is rotated away from the perpendicular, the center of the main lobe of the sonar beam is reflected off the feature in such a way that it is not returned to the receiver. Instead, a reflection from that part of the main lobe which is now perpendicular to the feature is returned to the receiver. Thus, the same range measurement is recorded as the sonar is rotated through an angle equal to the main lobe beam-width. The result of this is that a complete range-bearing scan is composed of a set of RCDs, one associated with each environment feature. Figure 5.2 shows a typical sonar scan together with a map of the room from which the scan was taken. A semi-circular RCD associated with each main environment feature, i.e. a wall, corner and edge, can be seen. Further insight into regions of constant depth and the parameters which govern them can be obtained by considering the impulse response of sonar as done in Appendix B.1.

Successive range returns constitute an RCD if the absolute range difference between the minimum and the maximum of the connected set of ranges is less than some ϵ_r. This is typically of the order of the

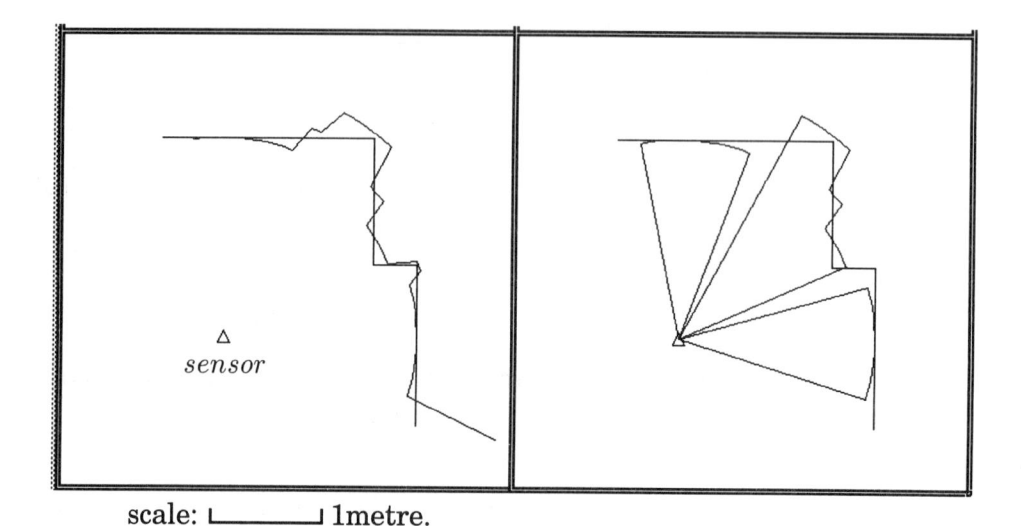

Figure 5.2: Sector scan showing (a) range data (b) RCDs extracted.

precision with which range can be measured. RCDs may be extracted from a range-bearing scan by simply segmenting range returns sequentially according to the criteria ϵ_r. The range of the RCD r_{RCD}, is given by the mode of the distribution of the ranges making up the RCD. We can define β_{RCD}, the angular width of the RCD, and β_{min}, the minimum acceptable angular width for what constitutes an RCD. This RCD description is illustrated in Figure 5.1 (b) for a symmetrical beam[3].

5.2.2 Sensing Limitations

Despite the improvements in data interpretation afforded by the RCD model, we can still identify the following shortcomings of sonar as used in robotics:

1. **Imprecise feature bearing.** Since the location of the actual target is given by the centre of the region of constant depth (RCD), the exact location of the target can only be determined from a scan which encompasses the bounds of the RCD. And so with a

---

[3]In practise the Polaroid beam is not symmetric and varies from device to device.

5.2 Understanding the Sensor

single range return, the criticism "Sonar has poor angular resolution ..."[82] still holds in the absence of a scan establishing the RCD edges. Scanning to establish RCD edges can be a significant handicap if precise orientation information is required quickly. In addition, RCD edges tend to be unstable as a result of range jumps induced by the thresholding nature of the receiving circuit and also because of partial feature obscuration. Moreover, RCD instability increases when there is relative motion (such as on a vehicle).

2. **Characterization of features.** The range and edge-angles defining a region of constant depth $(\theta_l, \theta_r, r_{RCD})$, give no indication as to the geometric nature of the target. The geometric classification of features is particularly important in map-building and navigation tasks. Unfortunately, a single standard time-of-flight sonar sensor does not *directly* provide information for the classification of features.

3. **Sonar data processing.** Once extracted, an RCD may be characterized by a single range value together with the bearing of its center (see Figure 5.1 (b)). However, obtaining this RCD information entails taking numerous sonar range readings which do not provide any additional useful additional information about the feature causing a reflection. This means that the quite considerable, albeit necessary, sonar data acquisition and processing does not actually yield a commensurate amount of information and it must be possible to do better. If, instead of taking a complete sonar scan of an environment, it were possible to take *single* range and bearing measurements, each corresponding to individual RCD reflectors, the number of range measurements required to generate a complete map of an environment could be reduced from several hundred (typical of scanning sonar systems) to about ten (the typical number of significant RCDs in a room). This would result in a considerable speed-up in data acquisition and consequently a reduction in data-processing.

In addressing these limitations, we are motivated by Leonard's orienteering analogy that; sensor action and attention should be focussed so as to measure only information that is directly relevant to the task at hand [136]. In order to achieve this, we describe a differential sonar sensor.

5.2.3 Differential Sonar Model

The use of two antennae measuring the same returned signal from two slightly different locations is termed *monopulse*. This principle is used extensively in tracking radar systems [195] where the difference signal measured is commonly called the monopulse signal. In radar the difference signal is constructed either from a comparison of the returned signal amplitudes (called amplitude-mono-pulsing) or from a comparison of the returned signal phase (called phase-mono-pulsing). A similar approach can be used with sonar assuming that we have devices which are capable of measuring amplitude and phase. This is not possible with the standard Polaroid device but can be achieved using modified sonar devices as shown by Barshan and Kuc [19]. In [124] a sonar sensor similar to the one described here is presented.

While not true phase-monopulse in the strict sense, the differential sonar model is analogous to the phase-monopulse technique in that it provides a measurement of the difference in time of arrival of a common signal at two spaced receivers. It is for this reason that we refer to the sensor as a *Differential Sonar*. The sensor consists of two standard Polaroid transducers mounted on a common baseline on a high-speed servo (or stepper motor) as shown in Figure 5.15. One of the transducers (T/R) both transmits and receives acoustic signals and the second transducer (R) only acts as a receiver. In operating the device, the following measurements are obtained:

1. A conventional *time-of-flight* range measurement is obtained from the T/R device.

2. The difference in time of arrival between the two receivers i.e. the *differential* denoted Δ. This is obtained by measuring the time between the reception on the first device to receive, *primary reception* and the time when the second device receives, *secondary reception*.

The principle is illustrated in Figure 5.3 and more details are given in Appendix B.

The form of the differential

Given that r is the range to the feature, d the base-line of the T/R and R devices and α the angle of inclination from the perpendicular to the

5.2 Understanding the Sensor

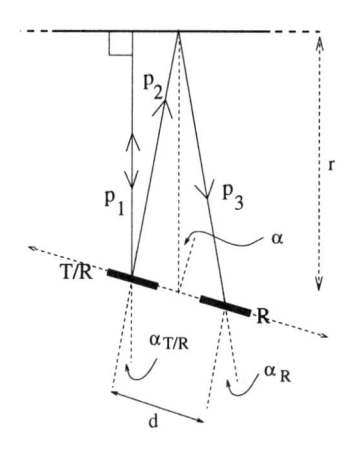

path length for T/R = $2p_1$

path length for R = $(p_2 + p_3)$

from geometrical considerations

$$2p_1 < (p_2 + p_3)$$

since TOF = f(path length)

TOF $_{2p_1}$ < TOF $_{(p_2 + p_3)}$

differential = TOF $_{2p_1}$ - TOF $_{(p_2 + p_3)}$

Figure 5.3: Illustrating the differential principle for a positive value of α. (See Appendix B.2 for a detailed analysis.).

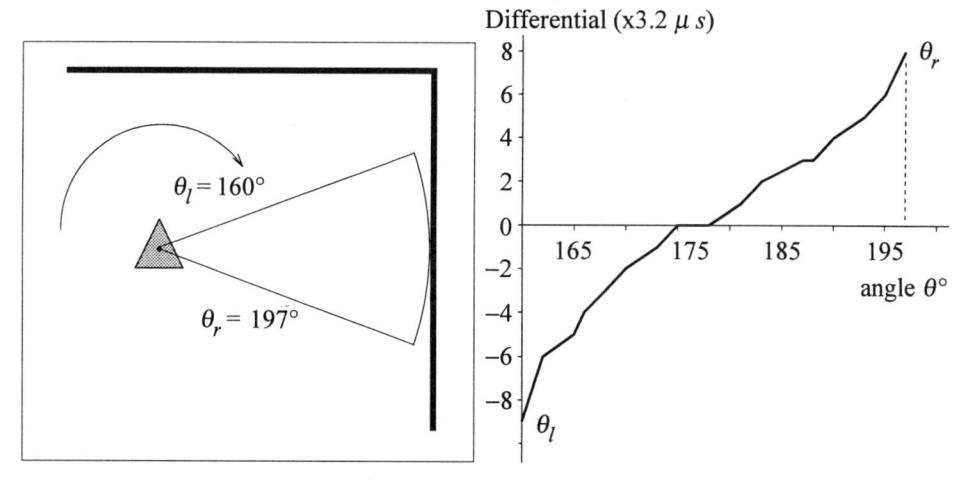

Figure 5.4: An extracted plane RCD and its associated differential. (a) shows the RCD and (b) the differential. The vertical scale in (b) is scaled by $3.2\mu s$ and the horizontal axis is the angular extent of the RCD in degrees.

plane of the feature, the general form of the differential time-of-flight is a straight line given by

$$\Delta \triangleq \frac{1}{c}\delta(r,d,\alpha), \tag{5.1}$$

where δ is the differential path length for primary and secondary reception as a function of r, d and α such that $-\alpha_{max} \le \alpha \le \alpha_{max}$. The precise form of the differential, or a given feature can be derived from geometrical considerations. The form of the differential path length δ for the features typically encountered in indoor environments is derived in Appendix B.2.

In Figure 5.4, an actual differential for a plane at a range of 3 metres in a controlled environment, is measured and plotted. It can be seen that when the target feature is centered between the receiver pair, the difference in signal arrival times is very close to zero, but not actually zero due to the offset induced by the small base-line. As the target moves away from center, the difference signal grows linearly with off-axis bearing. The exact bearing to the target feature can be determined by rotating the sonar pair so as to null the time difference signal and simply reading off the resulting aim bearing. If the feature moves, or more commonly in mobile robotics, the platform on which the sonar is mounted moves, the same signal can be used to maintain track on the target feature as shall be shown.

5.3 A Tracking Sonar Sensor

The Idea

With each return from the differential sonar, the following data is obtained

$$\text{return;} [r,\ \theta,\ |\Delta|,\ sign(\Delta)]. \tag{5.2}$$

Given that the return is within the bounds of an RCD (i.e. $\theta_l \le \theta \le \theta_r$), it can be seen from Figure 5.5 that θ is a computable distance ϵ away from the precise location of the target. The direction of the return away from θ_m is given by $sign(\Delta)$ (see Appendix B.4). An immediate consequence of this is that with a *single* return at an orientation θ within the width of the RCD, the precise orientation θ_m of a feature can be computed thus giving a bearing much more accurate than that obtained without use of the differential. This drastically reduces the angular (bearing) uncertainty. Therefore, differential information can be used to optimize

5.3 A Tracking Sonar Sensor

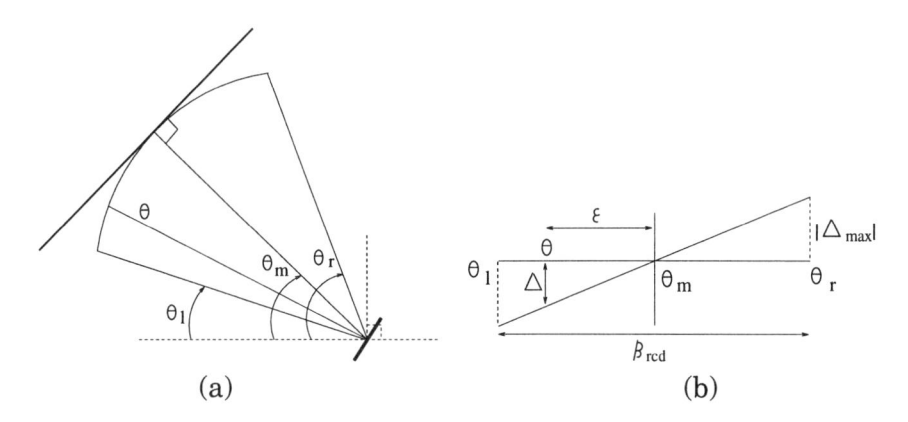

Figure 5.5: Information obtained from a return within the bounds of an RCD.

RCD extraction by reducing the number of range returns necessary to establish the RCD. This is done in the following way; once one edge of an RCD is detected during a scan, the precise bearing, and indeed location, of the corresponding feature can be predicted without requiring any additional measurements. More importantly, a simple tracking algorithm can be implemented which continually "corrects" for each non-zero differential by steering towards the zero (null) differential. This is the basis of the Tracking Sonar which gives a focus of attention capability.

In typical application environments, the measured differential is not as linear as predicted and the results of Figure 5.8 illustrate this for measurements taken in a typical cluttered indoor environment. So there is a need to develop a way of overcoming this uncertainty. This shall be addressed in Section 5.4.1. For our purposes here, we assume that we can obtain an estimate of the differential, i.e. a value for the differential which is less uncertain compared to the actual measured one. The estimated differential is used to implement the Tracking Sonar.

The Tracking Algorithm

As already stated, the Tracking Sonar works by continually correcting the aim bearing of the sensor towards the bearing corresponding to a zero differential and, by implication, the precise bearing of the feature. We define $s(k)$ as the magnitude of the motor motion in correcting the aim bearing of the sensor. This move step $s(k)$ can be determined by

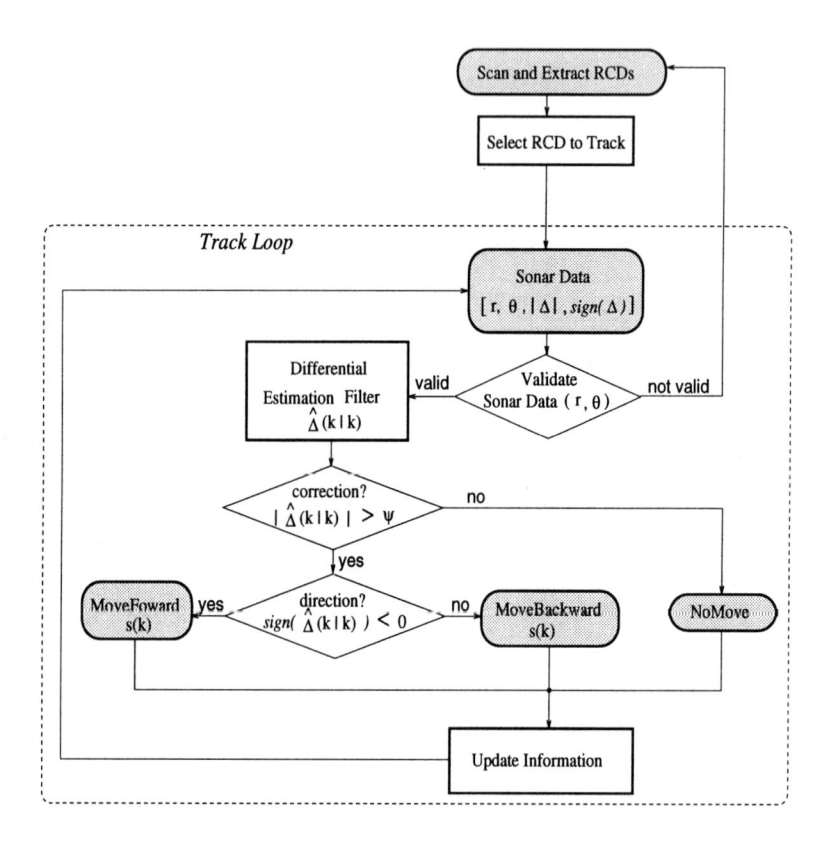

Figure 5.6: Tracking Sonar Algorithm. The shaded areas indicate hardware related procedures. Selection of the RCD to be tracked is application dependent and shall be discussed in the next Chapter.

5.3 A Tracking Sonar Sensor

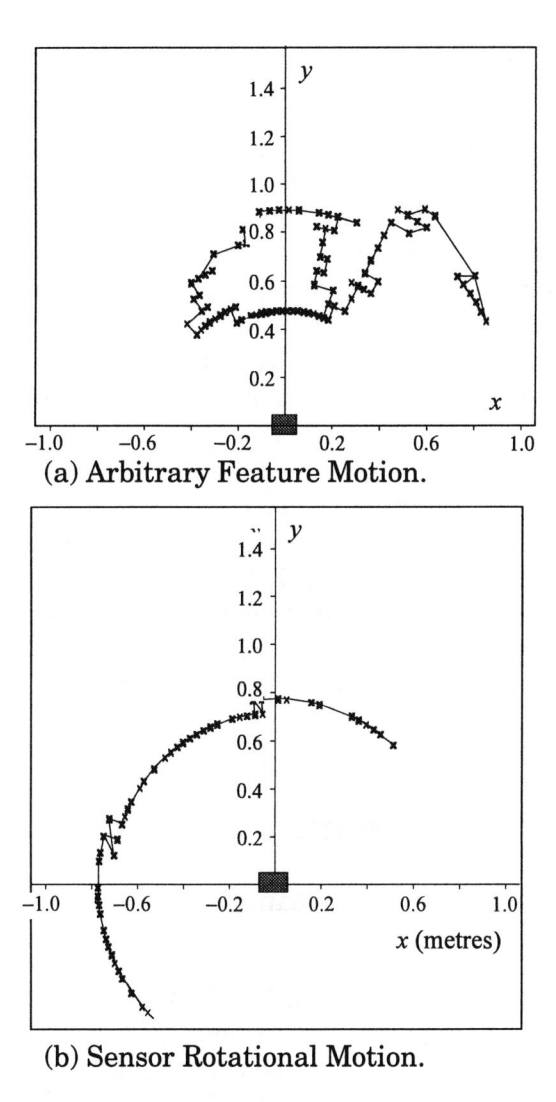

(a) Arbitrary Feature Motion.

(b) Sensor Rotational Motion.

Figure 5.7: Illustrating motion of a tracked feature in sensor ego-centric coordinates calculated from the range and orientation. In (a) the feature moves arbitrarily whilst the Tracking Sonar unit is stationary. In (b) the feature being tracked is stationary and the Tracking Sonar is rotating about a fixed point.

ϵ from the Figure 5.5, however, due to hardware limitations, better performance is obtained by correcting by a smaller step. The tracking algorithm is outlined in Figure 5.6, where $\hat{\Delta}(k \mid k)$ is the estimated differential (see Section 5.4.1). In a simple strategy, the range r is validated by a comparison with the current RCD range. If the error is acceptable, then the validated r becomes the new RCD range, otherwise, the range r is rejected resulting in a termination of the track. This simple approach can be improved upon with more detailed range and bearing validation incorporating checks for "spurious" measurements etc. The criteria for RCD selection depends on the application or goals of the system and shall be discussed later. The algorithm also includes a parameter ψ which defines the *dead-band* which is a minimum differential band within which correction for the differential error is deemed unnecessary.

Figure 5.7 shows some tracking results obtained using a differential sonar sensor implementing the above algorithm. Figure 5.7 (a) shows the motion of a feature which moves arbitrarily as observed by the sensor tracking it. In Figure 5.7 (b) the feature is stationary and the sensor platform rotates about its axis. The figure shows the relative positions of the feature as observed by the sensor using ego-centric coordinates. In all these results, the sonar data rates are approximately 30Hz.

5.4 Sensor Uncertainty and Probabilistic Model

5.4.1 Differential Uncertainty

The actual measured differential Δ incorporates some uncertainty and is sometimes spurious. This is evident from the measurements of the differential shown in Figure 5.8 for various features. To further highlight the uncertainty in the measurement of the differential, it is observed that when the differential sonar is fired repeatedly at a fixed orientation within the bounds of the RCD, the differential measured is not constant as expected. This is illustrated in Figure 5.9, where several hundred observations are taken under various environmental conditions (clutter, different feature materials, etc) and at various offsets within the angular extent of the RCD.

It is, therefore, evident that some smoothing or filtering operation is required to remove outliers and spurious readings. An approach

5.4 Sensor Uncertainty and Probabilistic Model

which is consistent with the other work presented thus far, is the use of estimation techniques. The actual measured differentials can be treated as noise corrupted observations of the true differentials, from which the true differentials can be estimated.

A Least Squares algorithm estimates an unknown *constant* parameter from uncertain measurements and so can be used in the case of sensor measurements whose values do not change with each sample. However, when using the Tracking Sonar, the parameter being estimated is time varying. The time-variant nature of the differential is due, firstly, to the algorithm continually correcting for error in aim bearing of the sensor and, secondly, to the possible relative motion between the sensor and the tracked feature. The correction motion of the servo can be considered as a control input which changes the state of the parameter being estimated. This obtains from the consideration that, the correcting motion of the motor towards the zero differential has the effect that a different value of the differential is measured (see Figure 5.5 (b)). The Kalman filter (whose information form is presented in Chapter 3) is well suited to the estimation of parameters which vary with each measurement.

Kalman filter for estimating the differential

The filter estimates the state $\mathbf{x}(k)$ and here the state is the differential Δ measured at time k. Since this state can be observed directly, the observation equation is given by

$$z(k) = x(k) + v(k), \tag{5.3}$$

where the observation noise is assumed to be additive, zero mean and Gaussian i.e.

$$v(k) = N(0, \sigma_\delta^2), \tag{5.4}$$

where σ_δ^2 denotes the variance associated with the differential measurement. Assuming a stationary feature, the changes in the true differential are only affected by a control input $u(k)$ to the motor. The control input $u(k)$, has the effect of reducing the differential by steering towards the center of the RCD. From the discussion in Section 5.3 (see Figure 5.5), the control is given by

$$u(k) = (-sign(\hat{\Delta}(k \mid k))) \times s(k) \times \mid \Delta_{max} \mid / \beta_{RCD}. \tag{5.5}$$

And so the state transition equation is

$$x(k) = x(k-1) + u(k-1) + w(k), \tag{5.6}$$

in which the transition noise is also additive and zero mean Gaussian i.e.

$$w(k) = N(0, \sigma_u^2)., \qquad (5.7)$$

where σ_u^2 is the variance associated with the state transition as effected by the control input $u(k)$. The resulting filter estimate $\hat{x}(k \mid k)$, is the estimated differential $\hat{\Delta}(k \mid k)$. (See Equations 3.52-3.56.).

This simple filter can be improved in several ways: A dynamic model of the feature can be introduced into the state transition Equation 5.6. In the formulation of the filter, it is possible to have a state which includes "velocity" as well as "acceleration". Such a larger state vector would improve the system's response to the dynamics of the differential in the tracking control loop by making available estimates of velocities and possibly accelerations. This is useful if one is interested in improving the response of the differential estimate to changes in the measured differential and in the control and damping of the motor. The exploration of these ideas is left to further research.

The differential estimates obtained using this simple filter with no control input $u(k)$ i.e. without the motor correcting the aim bearing towards a zero differential, are illustrated in the results of Figure 5.10. The results are obtained when there is no relative motion between the feature and the sensor, hence the measured differentials are similar to those of Figure 5.9. The estimated differential is in agreement with that predicted from the theory. Figure 5.11 illustrates the results obtained with a fully implemented algorithm which, in addition to estimating the differential, also steers the sensor towards the aim bearing corresponding to the precise bearing of the feature. The results in Figure 5.11 (a) show the differential (estimated and observed) while tracking a feature which moves at the indicated time-steps. Figure 5.11 (b) shows the results of the differential sonar tracking a stationary RCD target, while the sonar platform is rotated occasionally at the time steps as indicated. In these results we have set a dead-band ψ equal to 2 differential counts ($2 \times 3.2\mu$ s). This dead-band is set because the smallest correctional step possible, using the motor in the present prototype sensor hardware, corresponds to a differential of the order of 2 ($\times 3.2\mu s$). Consequently, a dead-band much smaller than 2 would result in continual correction regardless of relative motion, due to the lack of precision

5.4 Sensor Uncertainty and Probabilistic Model   147

in the positioning of the Polaroids by the motor. It must be emphasized that the dead-band is a consequence of hardware limitations. Similar hardware limitations can be identified in other sensor devices details of which must be incorporated into the sensor model.

5.4.2 Tracking Sonar Uncertainty

The Tracking Sonar provides measurements of range r and orientation θ data, where the orientation is referenced to a datum line on the sensor itself. The sensor can therefore be said to provide a 2-dimensional observation vector $\mathbf{z}$ given by

$$\mathbf{z} = [r, \theta]^T. \tag{5.8}$$

The range and orientation pair (r, θ), incorporates some uncertainty which can be modelled as follows: Assuming that the uncertainty in the measurements is in the form of additive noise, it can be written that

$$\mathbf{z} = \mathbf{z}_{true} + \mathbf{v}(\zeta, \mathbf{z}_{true}), \tag{5.9}$$

in which $\mathbf{z}_{true}$ is the true state being measured and $\mathbf{v}$ is the observation noise which is dependent on the true value $\mathbf{z}_{true}$, and on ζ the sensor operational parameters (differential filter parameters, dead-band etc). It will be noted that Equation 5.9 has been written in terms of the measured values themselves rather than a state $\mathbf{x}$ to avoid the need to state a model relating $\mathcal{X}$ and $\mathcal{Z}$, since this is application dependent. It can be assumed that $\mathbf{v} \in \mathcal{V}$, where $\mathcal{V}$ is a class of random variables [97]. It is, therefore, essential to develop a model of the additive noise $\mathbf{v}$ for a given range of sensor control parameters. It is usually assumed that the noise $\mathbf{v}$ is Gaussian, indeed the information filter derived in Chapter 3 makes such an assumption.

The Gaussian noise assumption is one which is made in a lot of cases without practical justification and validation. Often, justification is in the form of an appeal to the Central Limit Theorem[4]. For a sensor whose operation is determined by several operational parameters, it is important to validate practically the Gaussian assumption within the operational range of these parameters. A statistical model of the disturbance $\mathbf{v}$, can be developed by first considering the parameters

---
[4]The Central Limit Theorem states that for a sequence of independent identically distributed (i.i.d.) random variables y_i, the probability distribution function given by the summation $(1/\sqrt{n}) \sum^n y_i$ becomes Gaussian as $n \to \infty$.

A Sensor Model

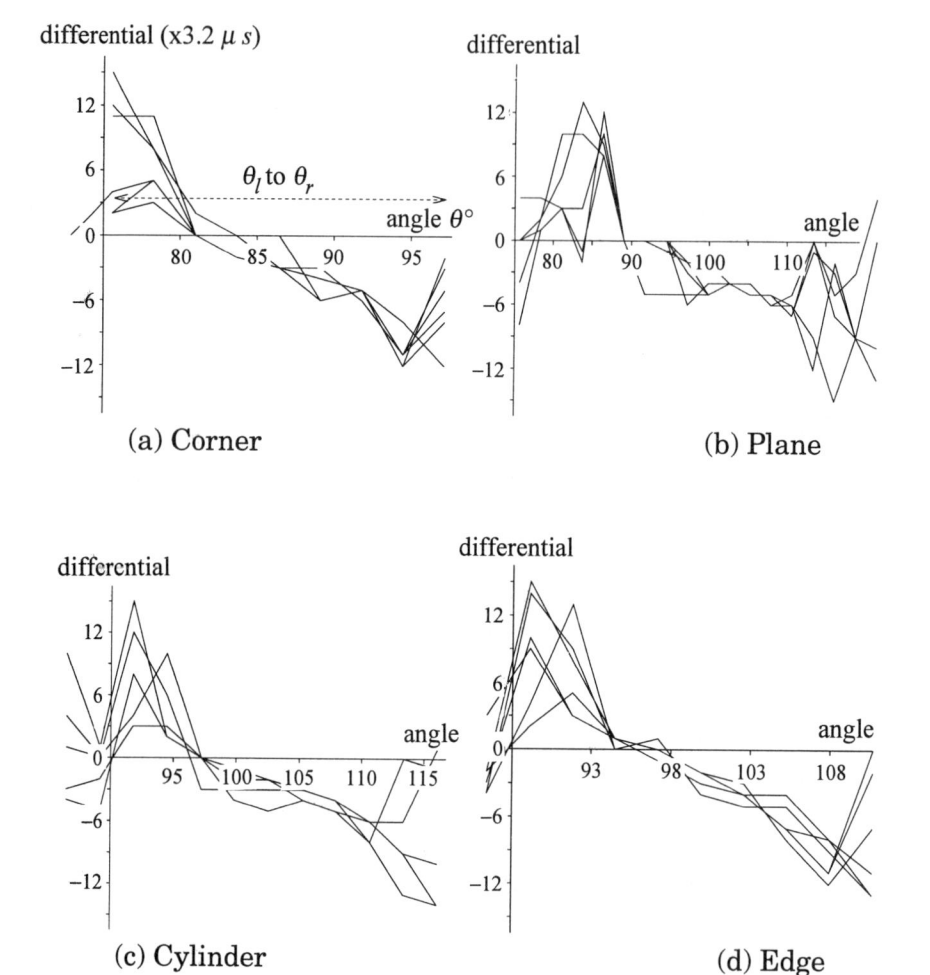

(a) Corner

(b) Plane

(c) Cylinder

(d) Edge

Figure 5.8: Plots showing multiple (5) differentials within the width of the RCD corresponding to each of the four target types. The angle axis is the orientation, from some fixed datum, of the ranges making up the RCD. Only differential information within the angular width of the RCD (θ_l - θ_r, see Figure 5.1) is of significance here. The vertical scale is the differential TOF $\times 3.2 \mu s$.

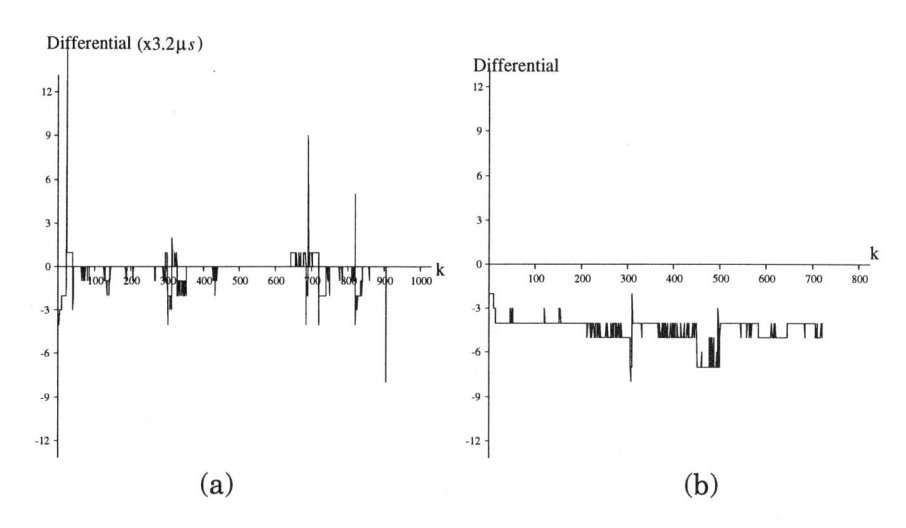

5.4 Sensor Uncertainty and Probabilistic Model

Figure 5.9: Differentials measured at a fixed orientation within the bounds of an RCD, highlighting the uncertainty associated with the differential. Theoretically, the differential should remain constant at some value depending on the degree of offset from the center or null position. In (a) there is no offset, in (b) offset is in negative direction.

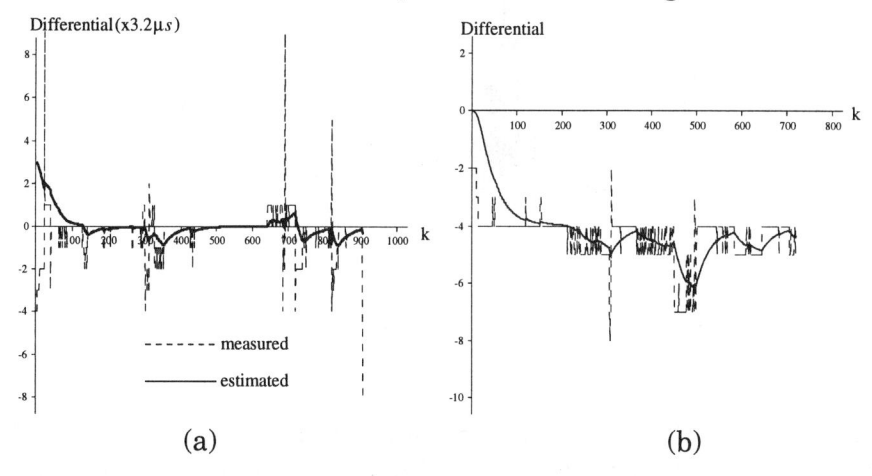

Figure 5.10: Plots (a) and (b) show the measured and the estimated differentials. In (a) the truncated predicted differential is 0, that is, no offset, in (b) the truncated predicted differential is -4. All values of the differential are scaled by $\times 3.2 \mu s$.

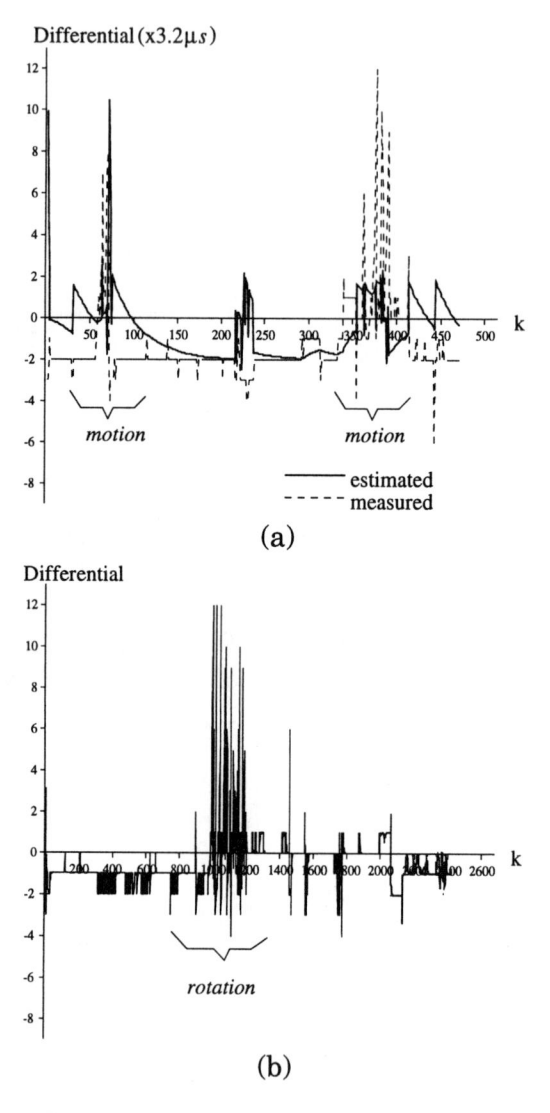

Figure 5.11: Results showing the measured and estimated differentials while tracking a feature. In (a), the differential sonar platform is stationary and the target is in motion from step 50 to 75 and from 350 to 390. In (b), the differential platform is rotating, while the feature is stationary. The platform rotates from step 1000 to 1200. In both (a) and (b) there is a no-correction differential dead-band of 2 ($\times 3.2\ \mu s$).

5.4 Sensor Uncertainty and Probabilistic Model

determining the performance and operation of the Tracking Sonar and the uncertainty in its observations. These parameters are as follows:

- **Differential estimator parameters.** A large value of differential observation noise σ_δ^2 (from the differential estimation filter, see Equation 5.4) is advisable when tracking features whose motion relative to the sensor is slow and deliberate. However, when motion is sudden and quick, what may appear as spikes in the differential may actually represent the sudden motion. In such situations, an observation noise model which assumes relatively little noise in the measurements results in better performance as it gives more credence to the actual observations themselves.

- **Hardware-related parameters** (e.g. Differential dead-band.) The dead-band is a consequence of hardware limitations, in particular limitations in positional accuracy and the minimum motion step size of the motor. In addition, electrical noise may result if there is a very high rate of correction motion by the motor, as occurs when every measurement leads to correction action, i.e. in the absence of a dead-band. This electrical noise may causes minor jitter in the positioning of the motor, giving unreliable differential measurements. This can be alleviated with reduced correction motion rates hence the dead-band and also with improved hardware designs.

A model should thus incorporate how the statistics of the information obtained from the sensor varies with the above factors. In the following section we introduce spectral estimation techniques which may be used to model the statistics of the measurement noise.

5.4.3 Use of Spectral Estimation Techniques

While the Tracking Sonar is tracking a feature, the sensor generates a stream of measurements. In analysing the measurements from the sensor, the range r and orientation θ over time can be treated as random processes. If there is no relative motion between the feature being tracked and the sensor, then the random processes can be viewed as stationary. When there is relative motion between the feature and the sensor, the processes then cease to be stationary. As an illustration of how to model information from the sensor, we shall only consider the stationary case, in other words, when there is no motion between the sensor and the feature being tracked[5][6]. Our interest in the proceeding analysis lies only in determining whether the process noise is white or not.

Random processes are defined in terms of their ensemble averages and these can be estimated. Our model shall be in terms of such averages. In practice, we require to estimate these averages from finite sequences. We consider a process y_k as realized (estimated) by the finite sequence $y(k)$, for $0 \leq k \leq N-1$. That $y(k)$ is an estimate of the random process y_k is made plausible by a consideration of ergodic processes. From $y(k)$, we can, therefore, estimate the averages for the process: The mean is estimated by

$$\hat{\mu}_y = \frac{1}{N} \sum_{k=0}^{N-1} y(k), \qquad (5.10)$$

and the variance by

$$\hat{\sigma}_y^2 = \frac{1}{N} \sum_{k=0}^{N-1} (y(k) - \hat{\mu}_y)^2. \qquad (5.11)$$

A *biased* estimate of the autocorrelation is given by

$$\tilde{\phi}_{\mathbf{yy}}(m) = \frac{1}{N} \sum_{k=0}^{N-|m|-1} y(k) \cdot y(k+m), \qquad (5.12)$$

with a variance of

$$\tilde{\sigma}_{\tilde{\phi}} = \frac{1}{\sqrt{N}} \hat{\sigma}_y^2, \qquad (5.13)$$

where $|m| < N$. Similarly, an *unbiased* autocorrelation is estimated by

$$\hat{\phi}_{\mathbf{yy}}(m) = \frac{1}{N-|m|} \sum_{k=0}^{N-|m|-1} y(k) \cdot y(k+m), \qquad (5.14)$$

---

[5]The assumption of stationary process when there is no motion is not strictly true, because the differential, which is used to maintain track, is essentially the result of a range measurement. Range measurements, even with no relative motion, can only be treated as stationary over very long periods because the measured range depends on the temperature of the intervening air which is usually treated as a random process having low frequency components (of the order of seconds). As a result, the closely timed range samples in the short interval are not really stationary. However, the error due to this is small enough to be ignored in our case.

[6]The case when there is motion requires that we consider the residuals of the process given an adequate process model, and attempt to describe these in terms of their averages by considering the residuals as a stationary process [118].

5.4 Sensor Uncertainty and Probabilistic Model

and the variance of the unbiased autocorrelation is given by

$$\hat{\sigma}_{\hat{\phi}} = \frac{\sqrt{N}}{N-|m|} \hat{\sigma}_y^2. \tag{5.15}$$

The power spectrum can be estimated using

$$\tilde{P}_N(\omega) = \sum_{m=-(N-1)}^{N-1} \tilde{\phi}_{yy}(m) e^{-j\omega m}, \tag{5.16}$$

for the biased autocorrelation and similarly for the unbiased one.

Some form of correction in the autocorrelation estimate is required for the low frequency trends due to the finite nature of the sequence used for estimation This can be done in the simplest cases by subtracting a constant term or by a differencing or filtering operation. The finite nature of N results in less damping in the autocorrelation than would be expected from theoretical considerations. Owing to this, Jenkins and Watts [118] draw the conclusion that it is sometimes dangerous to read too much into the visual appearance of the autocorrelation function especially from a short series. For this and other practical reasons [118][166] the estimated autocorrelation should realistically be used only as a guide to modelling or describing random processes.

For a truly zero-mean white process, the autocorrelation is large when $m = 0$, and is, ideally, zero elsewhere. The autocorrelation when $m = 0$ gives the variance of the process. For a finite sequence, basic tests for whiteness are as follows [118]:

- For a white process, the autocorrelation can be shown to be normal with a zero mean and a variance $1/N$, provided that N is large enough. Hence, any process whose autocorrelation approximates to this can be considered white. The degree of closeness to the true white process can be determined by considering statistical regions of confidence. For 99% confidence, the bounded region for whiteness is $\pm 2\sigma$ and for 95% the region is $\pm 3\sigma$, where σ is actually the estimated variance $\hat{\sigma}_y$. Oppenheim and Schafer [166] go as far as saying that as long as the normalized autocorrelation for $1 \leq m \leq 512$ is "much less" than the value at $m = 0$, then the random sequence can be considered as uncorrelated.

- In addition to the autocorrelation, the power spectra can be considered by taking the discrete Fourier transformation of the biased autocorrelation estimate i.e. $\mathcal{F}\{\tilde{\phi_{yy}}(m)/\tilde{\phi_{yy}}(0)\}$ for the biased autocorrelation. For a perfectly white sequence the power spectrum is minimally flat, i.e., equally distributed over all frequencies. Consequently, for a given power spectrum the degree of closeness to a flat spectrum can be used as a test for whiteness.

More sophisticated tests, based on Kolmogoroff-Smirnov [118] confidence regions can be done to complement these basic tests. An example of this is found in the work by Barshan for inertial sensors in [18].

5.4.4 Analysing Measurement Uncertainty

By considering the tracking of a stationary feature, a probabilistic model of the data can be developed by estimating and analysing the process averages as described in the previous section using finite sequences of the data. We consider the averages within ranges of operating parameters which give satisfactory performance of the sensor.

Figure 5.12 shows variations of orientation θ with differential observation noise σ_δ^2. The results show that for a stationary target the variance in the orientation of the target is reduced with a larger observation noise variance σ_δ^2. This is further illustrated in Table 5.1 for a fixed range and orientation. The table shows estimated values of the mean and variance of the range and orientation to the feature when there is no relative motion between the feature and the sensor. The results of Table 5.1 show that increasing the noise model eventually results in some instability. A compromise has to be reached between increasing the noise model and having a small enough model to give a good response to motion. And so, the value used for the noise models depends very much on the application and anticipated motion. With our particular hardware, it has been found that for a differential estimator process noise (see Equation 5.7) with variance fixed at $\sigma_u^2 = 0.002$, observation noise variances σ_δ^2 in the range 0.00001 - 0.1 give satisfactory tracking performance. The specific value in this range is determined by the nature of the features and the relative motion anticipated. Thus, the observation noise model (variance) is inversely proportional to the maximum angular speed of relative motion.

Analysis similar to that in Table 5.1 shows that with everything else the same, the estimated variances of r and θ increase with range. In addition, these variances also vary with the orientation angle θ in that, given that the field of view of the sensor is limited, the variance of the

5.4 Sensor Uncertainty and Probabilistic Model 155

(a) $\sigma_\delta^2 = 0.00001$

(b) $\sigma_\delta^2 = 0.001$

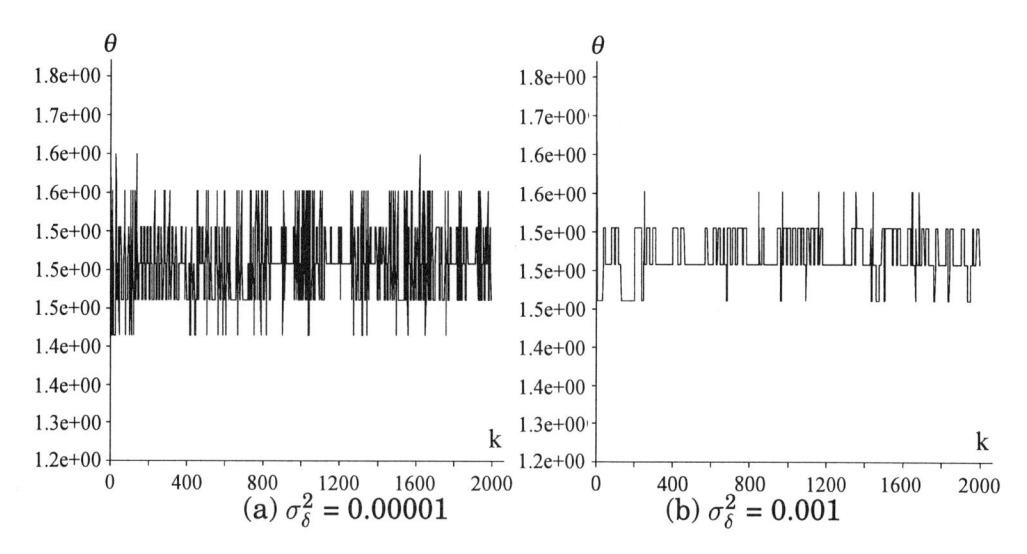

Figure 5.12: Plots showing the variation in bearing θ over time with no relative motion between the feature and the sensor for various differential estimator noise models σ_δ^2. With $\sigma_\delta^2 = 0.0000001$ (not shown) the feature was lost after about 500 steps. Variance of the process noise in the differential estimation filter is fixed at $\sigma_u^2 = 0.002$.

Table 5.1: Table showing typical values of the mean and variance of the range and orientation of the target with varying noise models at a fixed range and orientation. Value of the process noise ($\mathbf{Q}(k)$ in filter) fixed at $\sigma_u^2 = 0.002$.

Δ noise variance σ_δ^2	orientation mean $\bar{\theta}$ (rads)	θ variance σ_θ^2	range mean $\bar{r}$ (m)	range variance σ_r^2
0.00001	1.608	0.0033	1.513	0.0031
0.0001	1.605	0.0029	1.512	0.0016
0.001	1.621	0.0022	1.501	0.0007
0.01	1.606	0.0017	1.501	0.0005
0.1	1.613	0.0014	1.511	0.0005
0.5	1.601	0.0017	1.511	0.0004
1.0	1.617	0.0020	1.521	0.0005

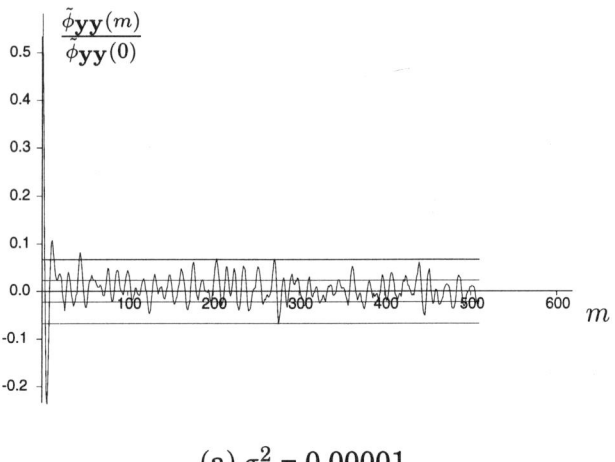

(a) $\sigma_\delta^2 = 0.00001$

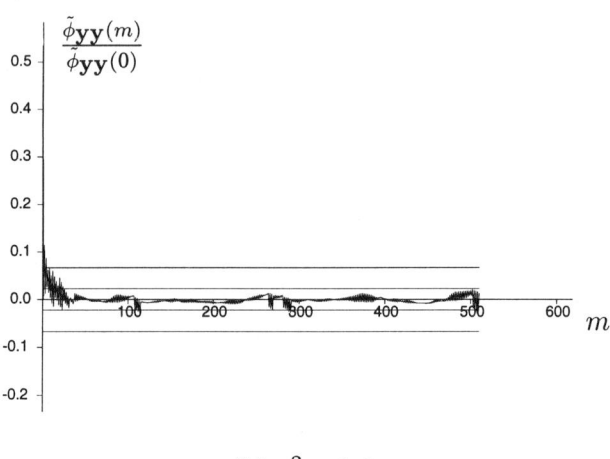

(b) $\sigma_\delta^2 = 0.1$

Figure 5.13: Autocorrelation estimates for the orientation during tracking of a stationary feature. The plots are for various values of the observation noise model, within the range giving satisfactory tracking performance. The autocorrelation is normalized, that is, $\tilde{\phi}_{yy}(0) = 1$. The plots also show $\pm 3\sigma^2$ and $\pm \sigma^2$ corresponding to whiteness confidences of 95% and 99% respectively.

5.4 Sensor Uncertainty and Probabilistic Model

(a) $\sigma_\delta^2 = 0.001$

(b) $\sigma_\delta^2 = 0.1$

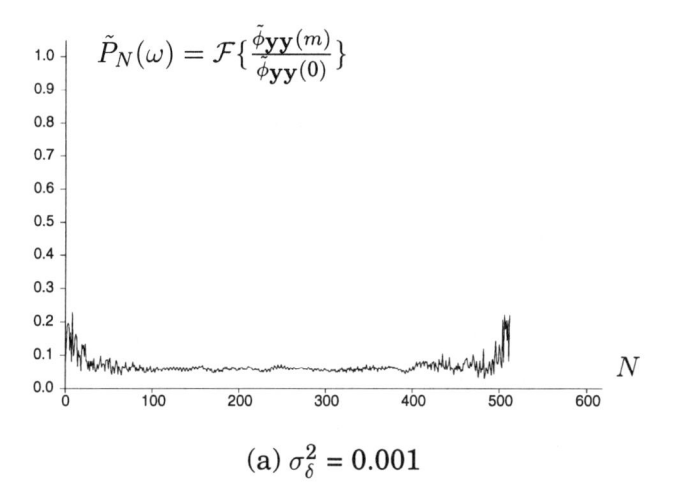

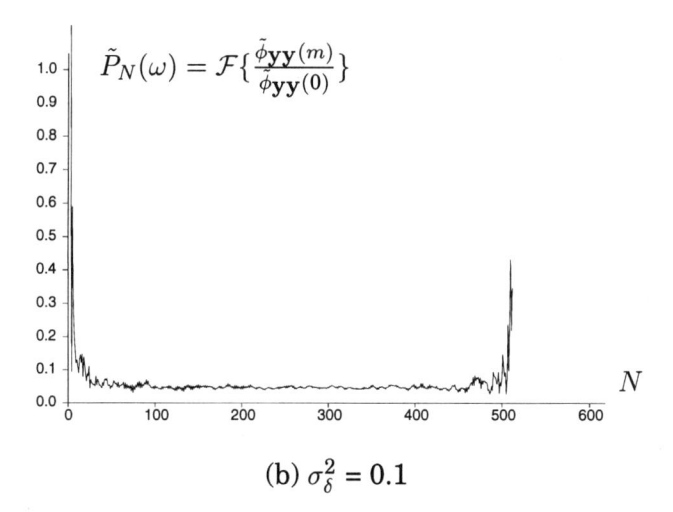

Figure 5.14: Power spectra using a Bartlett window $m = 512$ for the orientation during tracking for a stationary feature. The plots are for various observation noise models within the range giving satisfactory tracking performance.

orientation increases at the edges of the angular field of view making the variance somewhat dependent on the observation itself. Appendix B.2 gives functions which can be used to model this dependence of the variances on range and orientation.

For the ranges of the observation noise model σ_δ^2 and dead-band ψ which give satisfactory focus of attention performance, the data stream can be modelled as a true value corrupted by additive white noise. This can be justified by considering the autocorrelation and power spectra of the data streams under various conditions. By considering the extreme values of σ_δ^2 which give satisfactory performance, the estimated biased autocorrelation functions (Equation 5.12) for the orientation data can be plotted. From Figure 5.13 using the confidence tests described by Jenkins and Watts, it can be seen that the autocorrelation function is within the $\pm 3\sigma^2$ (for 95% confidence) region. Based on this together with the discussion by Oppenheim and Schafer [166] on the relative levels of the normalised autocorrelation function, we can justifiably model the orientation data as corrupted by white noise. We gain further confidence by considering the corresponding power spectra. Figure 5.14 shows the corresponding power spectra for the autocorrelation functions shown in Figure 5.13. These are approximately minimally flat as expected for a white sequence. It can also be noted from Figure 5.14 that as σ_δ^2 is increased, the modelling of the process as white becomes crude. However, bearing in mind the caveat by Jenkins and Watts concerning the need for correction at low frequencies in the autocorrelation, modelling the data stream as white seems justified.

The range data can be shown to exhibit similar noise characteristics to the orientation data, with the difference that in the range data, the variance is approximately an order of magnitude less than in the orientation data. This is as a result of the range validation on which continued tracking is dependent. This modelling of the data from the sensor has also been shown to hold for variations in the dead-band which show similar trends in the autocorrelation and power spectra.

5.4.5 Measurement Probabilistic Model

From the preceding analysis, the information supplied by the sensor can be modelled as having a mean about the true value ,with uncertainty due to additive white noise which has a variance that depends on both the measured quantities themselves and the operational parameters of the sensor. Thus we can write the components of the observation vector

5.4 Sensor Uncertainty and Probabilistic Model

of Equation 5.9, i.e. $\mathbf{z} = [r, \theta]^T$, as

$$r = r_{true} + v_r(\zeta, r_{true}),$$
$$\text{and} \quad \theta = \theta_{true} + v_\theta(\zeta, \theta_{true}), \quad (5.17)$$

where the noise is written as $v_r(\zeta, r_{true})$ to signify its dependence on the sensor control parameters ζ (differential filter parameters, dead-band etc) and on the true value (r_{true}) itself, and similarly for $v_\theta(\zeta, \theta_{true})$. Therefore, the noise vector $\mathbf{v}$ has a 2-by-2 covariance matrix $\mathbf{R}$ composed of the variances σ_r^2 and σ_θ^2 along the diagonal. The values of these variances can be determined for particular operational parameters as shown in the results presented (see Table 5.1). We assume that noise associated with the range measurement v_r is uncorrelated with the noise associated with the orientation v_θ. This assumption is made in the absence of analysis of the possible cross-correlations between v_r and v_θ at the same time-step.

Such a sensor probabilistic model is particularly useful because it facilitates the determination of the statistics of the data obtained. Consequently, measurement uncertainty can be taken into account quantitatively when making use of the information from the sensor. For a data fusion algorithm using a sensor with the model just described, the observation (measurement) noise covariance at the time step k (using the notation of Chapter 3) is written

$$\mathbf{R}[k, \mathbf{z}(k)] = \begin{bmatrix} \sigma_r^2(\zeta) \times g_r(r) & 0 \\ 0 & \sigma_\theta^2(\zeta) \times g_\theta(\theta) \end{bmatrix}. \quad (5.18)$$

By scaling the range and orientation variances, the functions $g_r(r)$ and $g_\theta(\theta)$ model the dependence of the variances on the measurements themselves. Details of the functions g_r and g_θ can be found in Appendix B.2. Such an observation dependent covariance $\mathbf{R}[k, \mathbf{z}(k)]$ is sometimes known as a *location covariance*.

Refining Models

In some sensor systems, it may turn out that the measurement noise $\mathbf{v}(\zeta, \mathbf{z}_{true})$ in Equation 5.9 is not white. One can proceed by refining the model such that the measurement noise becomes white. This can be done in the following manner; if $\mathbf{z}(k)$ is the actual measurement, and $\mathbf{z}_{model}(k)$ is the measurement predicted by the sensor model, the residuals denoted by $\epsilon(k)$ can be written

$$\varepsilon(k) = \mathbf{z}(k) - \mathbf{z}_{model}(k). \tag{5.19}$$

If the sensor model is good (in a predictive sense) then the residuals $\varepsilon(k)$ should be close to zero, or at least be a white noise process. If the residuals are not white, the model can be refined i.e.

$$\varepsilon(k) = \mathbf{z}(k) - \{\mathbf{z}_{model}(k) + \beta_1(k) + \cdots\}. \tag{5.20}$$

where the terms $\beta_i(k)$ are obtained by developing a more detailed model which explains the sensor measurements. Since

$$\{\mathbf{z}_{model}(k) + \beta_1(k) + \cdots\} = \mathbf{z}'_{model}(k),$$

this amounts to refining the sensor observation model. Conversely, the noise model $\mathbf{v}(\zeta, \mathbf{z}_{true})$ can itself be refined by separating out the correlated components and the white noise components, i.e.

$$\mathbf{v}(\zeta, \mathbf{z}_{true}) = \phi_v(k) + \mathbf{v}_{res}(k), \tag{5.21}$$

where $\phi_v(k)$ is the correlation model of the noise and $\mathbf{v}_{res}(k)$ is the residual white noise. The choice between these two approaches is dependent on the practical ease of each which in turn depends on the sensor. Schwartz and Shaw [189] present methods that can be used to develop the above.

5.5 Discussion

5.5.1 Summary of Performance and Limitations

As a complete sensor, the differential sonar, comprising the Polaroids, driver electronics and servo can be operated in several modes. Its use in extracting RCDs is similar to the work which has been described in the literature, such as in [134][113]. In this area and in light of the limitation of sonar discussed in Section 5.2.2, the differential sonar offers the improvement that the RCD extraction algorithm can be optimized using the availability of centering information from the differential as discussed in Section 5.3. The differential sonar, as implemented here, does not provide information which can be used directly to classify the feature. However, as an indirect consequence of the ability to fixate on

5.5 Discussion

a given feature, a method based on the classification algorithm can be developed which results in classification of the feature being tracked. In [124], Kleeman and Kuc describe a similar sensor making use of two transmitters and receivers which has the advantage that measurements can be obtained that can be used directly to differentiate various geometrical features.

The use of the differential sonar in its tracking (focus of attention) mode has interesting implications for data fusion and sensor management. The performance in this mode is discussed in Appendix B.3 and summarized here.

1. **Speed of relative motion.** This is the speed of relative motion between the tracked feature and the Tracking Sonar. The maximum speed of relative motion is determined by the sonar firing rates, and the range validation scheme used. Nominal values of the speed are of the order of 0.6 radians per second for the angular component and 2.5 metres per second for the linear component at firing rates of up to 30Hz.

2. **Tracked feature.** The nature and range of the tracked feature determine the performance of the Tracking Sonar. Performance improves with linearity in the differential characteristic and decreases with range. Features consisting of planes, corners, edges and cylinders are tracked well at a range of up to 5m.

3. **Feature discrimination.** This is the ability to distinguish between features which nearly coincide spatially. For this, a more stringent measurement validation scheme together with increased measurement accuracy are required.

4. **Multi-sensor operation.** The Tracking Sonar is able to function in an environment where other Tracking Sonars are operating, albeit with slightly reduced performance.

Given the above factors, the operational parameters of the Tracking Sonar can be optimized to give the best results depending on the anticipated relative motion, nature of tracked feature, operating range and hardware capabilities. In the probabilistic model, the sensor parameters ζ (i.e. differential estimator parameters, dead-band etc), are the significant factors determining the variances of the measured quantities r and θ. In addition, it is noted that the measured quantities themselves are also a factor. And so a complete model should describe the dependence of the variances of r and θ with the measured quantities themselves. Appendix B.3 gives an empirical approximation of this dependence.

Insight into the setting of the parameters ζ is built with experience and consideration of some of the variations discussed in the preceding sections. The results of Section 5.4.4 suggest possible settings for various filter and other parameters. There is ample room for further development of the hardware itself and optimization for various scenarios and applications. We leave the improvement of this technique to further research.

5.5.2 A Modular Data Fusion Sensor

The Tracking Sonar is packaged in a modular fashion, based on a generic decentralized architecture that has been developed at Oxford. The architecture provides the requisite processing power using a T805 Transputer with full floating point ALU. Communication links are available for connection to other similar sensing nodes. Appendix B.4 summarizes the architecture. Figure 5.15 shows a complete Tracking Sonar based sensing node built using this architecture.

Such a modular implementation of the Tracking Sonar has a significant impact on the way multi-sensor fusion algorithms, for vehicle applications in particular, can be realized:

1. The sensor architecture is fully autonomous and, as such, well suited to the distributed and decentralized **multi-sensor architectures** developed in Chapter 3. The availability of processing power means that complex algorithms can be implemented *locally* on each sensor node.

2. The ability of each Tracking Sonar to focus attention on a given feature means that in a decentralized system consisting of such sensors, each one can focus on a particular aspect of the problem and all the results fused to give a global result. This makes the sensor an ideal candidate for the implementation of the **data fusion algorithms** developed in Chapter 3.

3. The ability to focus attention only on a selected feature presupposes ability to decide on features to focus on. This presents an

5.6 Bibliographical Note

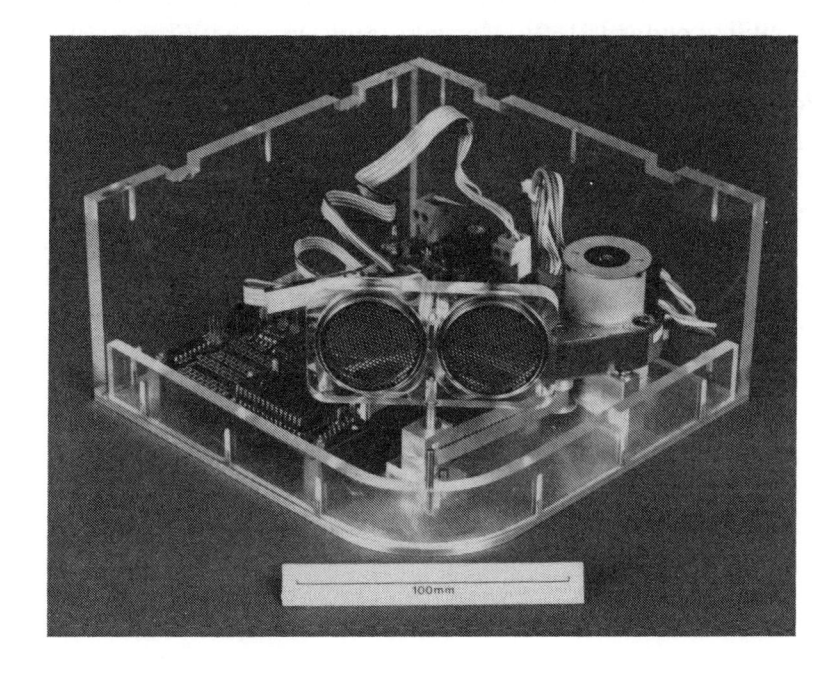

Figure 5.15: A completed modular Tracking Sonar sensing node.

ideal **sensor management** application requiring the making of decisions, given possibly several alternatives.

5.6 Bibliographical Note

Various other approaches to sensor models abound in the literature. For comparison, reference can be made to those presented by Dario *et al* [61], Flynn [82], Hackwood *et al* [93], Barshan and Kuc [19] and Barshan and Durrant-Whyte [18, 17], Hovanessian [111], McKendall and Mintz [150] which range from physical descriptions of how sensor work to statistical models of uncertainty. Approaches to sensor modelling which are different to the one we have taken are latent in the literature and references are given in Chapter 1. Models for specific sensor devices which have been discussed by other researchers include radar [25][99], infra-red [82][31], inertial sensors [18] and those mentioned in the survey by Luo [140].

Models for interpreting sonar data have been discussed by Kuc and

Seigel [127] from which we draw considerably and also by Elfes [74], Hallam [98], Leonard [137], Manyika and Durrant-Whyte [147], Barshan and Kuc [19]. Work describing the use of sonar in mobile robotics includes that by Elfes [73, 74], Kuc [125], Manyika and Durrant-Whyte [147], Moravec and Elfes [156], Hu and Probert [113], Brady *et al* [33], further references are also provided in Chapter 6. Other references in robotics literature to sensors with a focus of attention capability are in vision systems, as exemplified in the work of Andersson [8], Murray *et al* [192], and that by Eklundh *et al* [168], and also in tracking radar systems [195].

6 Data Fusion for Robot Navigation

Information can tell us everything. It has all the answers. But often they are answers to questions we have not asked, and which doubtless do not even arise.

- Jean Baudrillard

Implemented data fusion system are capable of obtaining copious quantities of information. But most of this information is often not used, simply because it is not relevant to the problem at hand. But by directing sensing efforts, it is possible to ensure that only information which provides answers to 'questions asked' is obtained. This results in more efficient processing of information to obtain the knowledge sought. Whilst certainly not a general panacea, this principle can be applied to a wide variety of systems. The work in this chapter demonstrates the principle by applying the algorithms presented thus far in conjunction with the sensor model of the preceding chapter to, not unexpectedly, robot navigation.

6.1 Mobile Robot Navigation

Mobile robotics research has been dominated by work aimed at developing a competence for navigation and guidance. There are several aspects of this to be considered; firstly, low-level capabilities of *obstacle avoidance*, *localization* and *map-building*, and secondly, higher-level competences for *path planning* and *task planning*. The higher level competences presuppose the lower level ones. The work outlined in this book thus far can be used to develop a competence for localization and to facilitate map-building. Localization is defined as the process of answering the question *"where am I?"*[137][66]. The basis of most localization techniques is the use of beacons with respect to which location can be computed [54][32][1]. Put in the context of the ideas presented in this book, localization is concerned with the gain of information about a state x, as in Equation 2.4, where x is the location of the robot vehicle estimated on the basis of observations $\{Z^k\}$ of beacons.

The process of localization and, indeed, navigation is well understood in maritime and aerospace systems. Since the dawn of antiquity, navigation has depended upon the observation of known and reliable beacons such as landmarks and the positions of heavenly bodies. In applying this method to autonomous robot navigation, difficulties are encountered owing to the use of uncertain sensor observations to correlate or maintain correspondence with navigational beacons from which location is estimated. The problem is exacerbated when attempts are made to use features that occur naturally in the environment as beacons. The initial difficulties here arise as follows; firstly, determining the usefulness (informativeness) of a naturally occurring feature as a beacon is difficult, particularly, when there is no *a priori* information available concerning the feature. Secondly, naturally occurring features may be of a form that is not unambiguously distinct, as observed by a sensor, in such a way that they may be used as beacons. These, seemingly insurmountable, difficulties can be significantly reduced if a map consisting of naturally occurring features is provided. The utility of such a map is assured if the descriptions of features in the map are in a form that is consistent with how a sensor would view such features. This requires sensor-based descriptions of features occurring naturally in the robot's environment. The work by Leonard and Durrant-Whyte [137] is representative of autonomous robot navigation using naturally occurring features as beacons.

Localization can be achieved using sonar and an algorithm based on the standard centralized EKF which matches extracted feature (RCD) information with a given map and estimates vehicle location on the basis of the successful matches. The matching process is done every processing cycle. The net result of this approach is that the vehicle executes a "move-stop-look" sequence with considerable time being spent extracting RCDs and matching them with the map[2]. In suggesting improvements, Leonard states;

---

[1] There are several alternatives to this such as the use of *occupancy grids* described by Moravec [155] and Elfes [74].

[2] Leonard describes using high density scans taking up to 2 minutes to obtain. He reports location estimates taking about 1 second to obtain [134].

> "For fast localization we want to build a system that maintains *continuous* map contact, ... The power behind this approach is that when correspondence can be assured, by "locking on" to a target, a high bandwidth stream of easily usable [correctly associated] information becomes available ... We envision a robot which maintains *continuous map contact* "grabbing hold" of corners, planes and cylinders in the environment, using them as hand rails ... by maintaining continual correspondence with some subset of map features perception can be put in the loop to provide high bandwidth position control." [134].

Assuming this approach, our starting point is the Tracking Sonar, developed in the previous chapter, which has the ability to focus attention on a given geometrical feature such as the ones naturally occurring in indoor environments. Having established correspondence with a known feature in an *a priori* map, a focus of attention sensor provides a high bandwidth of localization estimates for as long as focus of attention is maintained. Based on these concepts, this chapter outlines;

1. distributed localization using focus of attention sensors and

2. the distributed classification of features.

6.2 Estimation of Location

6.2.1 Sensor-based Feature Descriptions

As a preamble to the localization scheme, we define geometric models for features occurring naturally in the environment which can be used as beacons. The same basic features that were considered when analysing the differential, i.e. planes, corners, edges and cylinders, are assumed. The features are illustrated in Figure 6.1.

1. **Plane.** A plane is defined by

$$t_p = (p_{norm}, p_\phi, p_v), \tag{6.1}$$

where p_{norm} is the length of the normal from the plane to the Cartesian coordinate origin of the environment, p_ϕ is the bearing of this normal and p_v defines which side the plane is visible from.

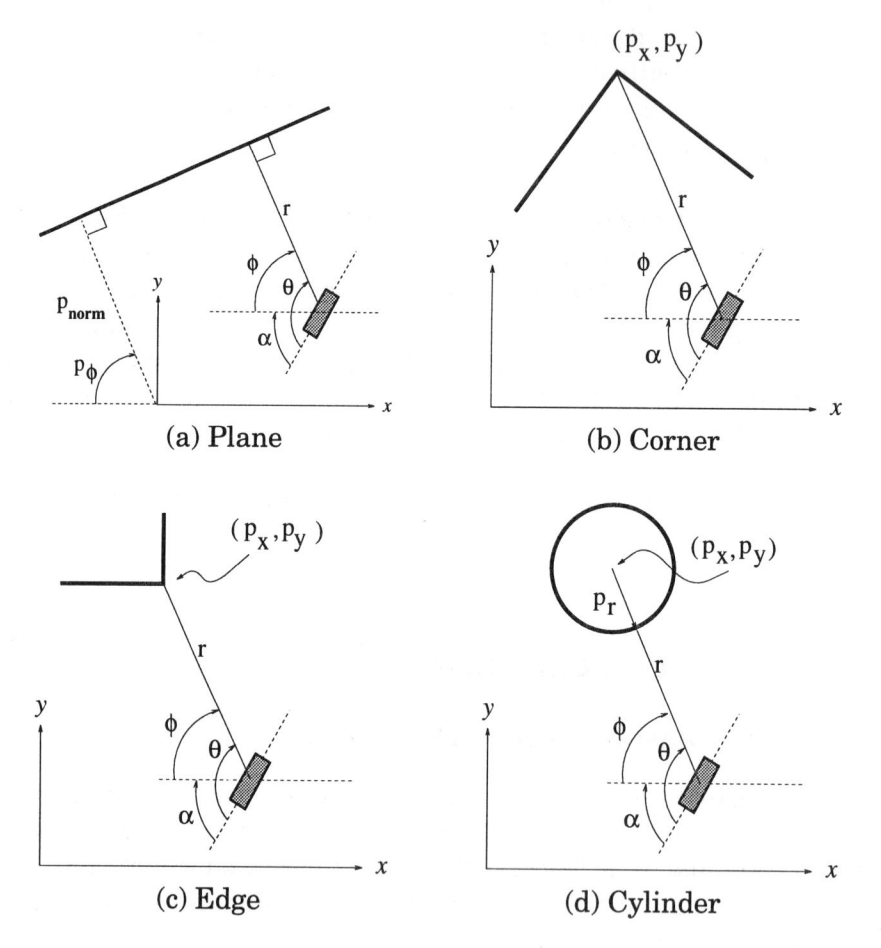

Figure 6.1: Sensor-based feature descriptions in a global coordinate system.

6.2 Estimation of Location

2. **Corner.** A corner is defined by
$$t_c = (p_x, p_y), \qquad (6.2)$$
where p_x and p_y are the coordinates of the location of the point of intersection of the planes making up the corner. It is assumed that the corner is right-angled.

3. **Edge.** An edge is defined by
$$t_e = (p_x, p_y), \qquad (6.3)$$
where p_x and p_y are the coordinates of the location of the point of intersection of the planes making up the edge.

4. **Cylinder.** A cylinder is defined by
$$t_{cyl} = (p_x, p_y, p_r), \qquad (6.4)$$
where p_x and p_y are the coordinates of the center of the cylinder and p_r is the radius of the cylinder.

In the above, it can be argued that the frame of reference of the coordinate system is unimportant on the basis that, from a functional perspective, only the relative spatial locations are of significance. Certainly, it is possible to use an egocentric (sensor platform) coordinate system. But, clearly, such referencing is, in practice, determined by the application and the nature of the environment information available.

6.2.2 Localization Algorithm

The algorithm for estimating location is a recursive one in which information about the location is updated based on the occurrence of observations that contain information about the location. Such an algorithm has already been presented, to wit, the information filter of Chapter 3. What is required now is a description of how the observation of features, such as planes, corners, cylinders and edges, relates to the location of the sensor making the observation. We define the following:

State

The location of the sensor platform in the environment is defined as the state
$$\mathbf{x}(k) = [x(k), y(k), \alpha(k)]^T, \qquad (6.5)$$

where $x(k)$ and $y(k)$ are the Cartesian coordinates of the platform location and $\alpha(k)$ is the orientation at time k.

Observations

The Tracking Sonar makes an observation of the range r and the bearing θ to a feature t_a. This pair (r, θ) at time k defines the observation vector $\mathbf{z}(k)$. The observation vector is a function of the state and the observed feature. Thus we can write that

$$\mathbf{z}(k) = [r(k), \theta(k)]^T = \mathbf{h}[k, t_a, \mathbf{x}(k)] + \mathbf{v}(k), \quad (6.6)$$

where $\mathbf{h}[\cdot]$ is the observation model and $\mathbf{v}(k)$ the associated observation noise. The observation model can be described geometrically for each feature type. From Figure 6.1 (a), the observation model for a plane is given by

$$\mathbf{h}[k, t_p, \mathbf{x}(k)] = \begin{bmatrix} p_{norm} - x(k)\cos(p_\phi) - y(k)\sin(p_\phi) \\ p_\phi - \alpha(k) \end{bmatrix}, \quad (6.7)$$

and that for a corner is given by

$$\mathbf{h}[k, t_c, \mathbf{x}(k)] = \begin{bmatrix} \sqrt{(p_x - x(k))^2 + (p_y - y(k))^2} \\ \tan^{-1}\left[\frac{p_y - y(k)}{p_x - x(k)}\right] - \alpha(k) \end{bmatrix}. \quad (6.8)$$

Since $t_c = t_e$, we have that; $\mathbf{h}[k, t_c, \mathbf{x}(k)] = \mathbf{h}[k, t_e, \mathbf{x}(k)]$. The observation model for a cylinder is given by

$$\mathbf{h}[k, t_{cyl}, \mathbf{x}(k)] = \begin{bmatrix} \sqrt{(p_x - x(k))^2 + (p_y - y(k))^2} - p_r \\ \tan^{-1}\left[\frac{p_y - y(k)}{p_x - x(k)}\right] - \alpha(k) \end{bmatrix}. \quad (6.9)$$

It will be noted in these observation descriptions that, unlike the algorithm of Leonard [134], use is made here of *both* the range and the orientation in the localization cycle. Leonard's reason for not using orientation is the uncertainty associated with the orientation when using the standard Polaroid device. This is overcome by the fact that, in addition to range, accurate orientation information with a known noise model can be obtained.

As observed in Chapter 5 for the Tracking Sonar, the variances associated with the observations vary in magnitude with the observations themselves. Owing to this, the covariance matrix of the observation noise is stated more accurately as $\mathbf{R}[k, \mathbf{z}(k)]$.

6.2 Estimation of Location

State Transition

The changes in the location of the sensor platform (vehicle) can be described by a state transition equation. The state transition from time k to time $(k+1)$ is given by

$$\mathbf{x}(k+1) = \mathbf{f}[k, \mathbf{x}(k), \mathbf{u}(k)] + \mathbf{w}(k), \qquad (6.10)$$

where $\mathbf{w}(k)$ is the state transition noise with covariance matrix $\mathbf{Q}(k)$. The control input $\mathbf{u}(k)$ represents the motion of the sensor platform. Various methods have been proposed to model the kinematics of robot vehicles some of which are quite complex [196][4]. A computationally simple and useful method is the one based on a point kinematic model. In this model, every motion can be represented by a rotational component and a linear component. And so, the control input is written

$$\mathbf{u}(k) = [T(k), \Delta\alpha(k)]^T, \qquad (6.11)$$

where $T(k)$ is the linear component of the motion at time-step k and $\Delta\alpha(k)$ is the change in orientation at time-step k and thus represents the rotational component of the vehicle motion. The transition model can be written

$$\mathbf{f}[k, \mathbf{x}(k), \mathbf{u}(k)] = \begin{bmatrix} x(k) + T(k)\cos(\alpha(k)) \\ y(k) + T(k)\sin(\alpha(k)) \\ \alpha(k) + \Delta\alpha(k) \end{bmatrix}. \qquad (6.12)$$

Estimation

Estimates of the sensor platform location are obtained using the nonlinear information filter outlined in Chapter 3, where Equation 3.56 gives the estimate $\hat{\mathbf{x}}(k \mid k)$.

6.2.3 Implementation

Assumption 3 : *The environment is a static two-dimensional world consisting of the feature types described in Figure 6.1.*

This assumption follows from the way our sonar sensing model has been developed[3]. The features are expected to be static because the

---

[3]Extension to the 3-dimensional case is possible but is outside the scope of the monograph. This is because for localization, as defined here, location in the vertical plane provides no useful information given that the robot vehicle only moves in the horizontal plane.

algorithm does not include target dynamics. However, the algorithm can be extended to include non-static features and a dynamic map.

Assumption 4 : *Available is an accurate map from which predicted observations can be generated and a corresponding facility for matching actual observations with the predicted ones.*

After extracting RCDs, they are matched with those predicted from the environment map. Predicted RCDs are generated from a line-segment map of the environment using the sonar model of Chapter 5 and the location of the vehicle [127][134]. The matching algorithm can simply be based on a validation gate. More complex methods exist such as those based on the Mahalanobis distance [76] and Nearest Neighbour Association [177]. Nearest neighbour association becomes unreliable with sonar data because of the occurrence of spurious false ranges which can falsify matches. However, if a sophisticated validation method is utilised in the extraction of RCDs which takes account of spurious range measurements, nearest neighbour methods can then be effectively used. Yet more sophisticated probabilistic data association methods such as the PDAF and JPDAF [45] have been used in other sensor applications. But the underlying assumption in these probabilistic approaches that the false observations of RCDs will be randomly distributed can not always be satisfied with acoustic data thus rendering these methods in their standard formulation unsuitable. An alternative approach is to not make immediate decisions about matches but defer whilst maintaining multiple vehicle location hypotheses which can then be tested in the light of subsequent observations. Such an approach can make use of a Multi-Hypothesis Tracking (MHT) filter.

In keeping with other work presented in this book, we can use a matching algorithm based on *innovation* sequences [16] and the use of a validation gate. For each predicted observation $\hat{\mathbf{z}}_{t_a}$ associated with feature t_a and for each observation $\mathbf{z}_i$ (i.e. extracted RCD), such a validation gate is given by

$$\nu_{it_a}^T(k)\mathbf{S}_{it_a}^{-1}(k)\nu_{it_a}(k) \leq \gamma, \qquad (6.13)$$

where ν_{it_a} is the innovation sequence obtained from

$$\begin{aligned}\nu_{it_a}(k) &= \mathbf{z}_i(k) - \hat{\mathbf{z}}_{t_a} \\ &= \mathbf{z}_i(k) - \mathbf{h}[k, t_a, \hat{\mathbf{x}}(k \mid k-1)],\end{aligned} \qquad (6.14)$$

6.2 Estimation of Location

and $\mathbf{S}_{it_a}(k)$ is the innovation covariance obtained from

$$\begin{aligned}\mathbf{S}_{it_a}(k) &= E\left\{\nu_{it_a}(k)\nu_{it_a}(k)^T\right\} \\ &= \nabla_{\mathbf{x}}\mathbf{h}[k, t_a, \hat{\mathbf{x}}(k \mid k-1)]\mathbf{P}(k \mid k-1)\nabla_{\mathbf{x}}\mathbf{h}[k, t_a, \hat{\mathbf{x}}(k \mid k-1)]^T \\ &\quad + \mathbf{R}_i(k),\end{aligned} \quad (6.15)$$

and γ is a validation constant.

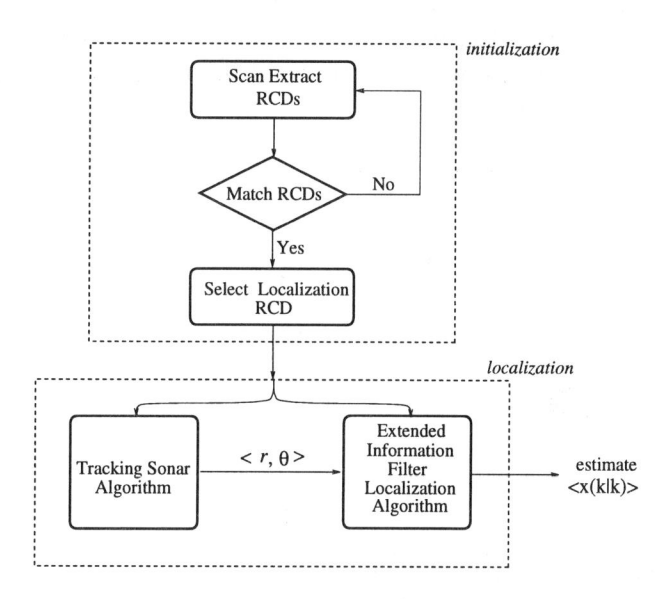

Figure 6.2: Localization algorithm. The Tracking Sonar and Localization algorithms are executed in parallel.

The localization algorithm proceeds as illustrated in Figure 6.2. If after scanning no matches are made, the scan and extraction process is repeated. If several matches are made, one is selected based on some criteria. In a simple strategy the best matched RCD, i.e. the match with the smallest innovation (see Equation 6.13), is selected. Other selection mechanisms include selection based on a marginal information metric or that based on bargaining aimed at obtaining a Bayes group action (see Chapter 4). The parameters of the selected RCD can be used to initialize the information filter in addition to being used to initialize the Tracking Sonar algorithm (see Figure 5.6). The Tracking Sonar algorithm and the information filter are then executed in parallel. Of significance is the fact that, the state estimates $\hat{\mathbf{x}}(k \mid k)$ are obtained at the sensor measurement rate. The parallel execution terminates when the tracking algorithm signals that the feature has been lost, following which the procedure outlined in Figure 6.2 can be repeated.

6.2.4 Results from estimating the location of a sensor platform

The following results illustrate the performance of the localization algorithm as implemented on a single sensor. The estimates shown in all the results that follow were obtained in real-time at a data rate of approximately 30Hz.

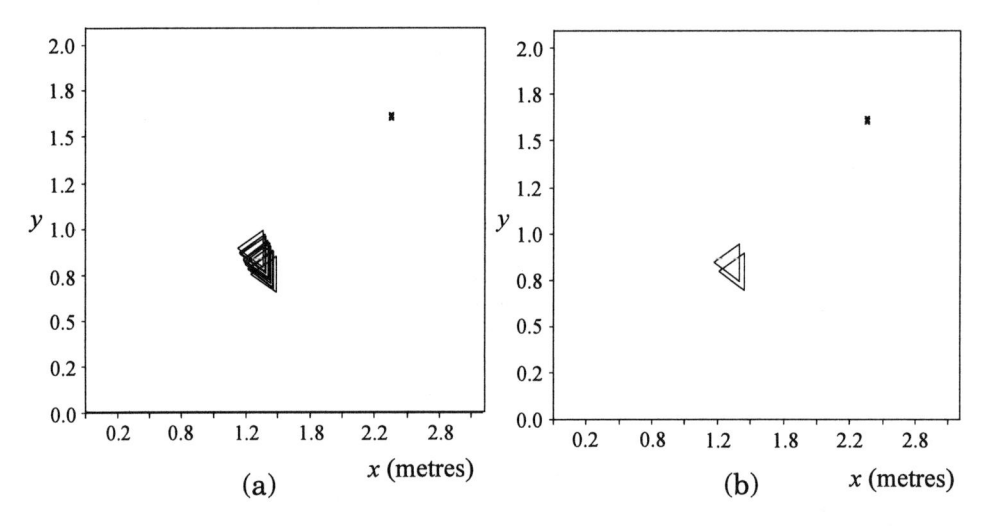

Figure 6.3: Localization estimates in global coordinates for a stationary sensor platform using various observation noise models. In the figures, positional changes of at least 0.04m are noted. Values in $\mathbf{R}(k)$ are set as follows; (a) $\sigma_r^2 = \sigma_\theta^2 = 0.0001$, (b) $\sigma_r^2 = 0.01$, $\sigma_\theta = 0.1$. The transition noise matrix $\mathbf{Q}(k)$ is a fixed diagonal matrix with the diagonal elements set to 0.001. The position "X" marks the location of the feature used for localization.

In the first instance, the localization results for a stationary sensor platform for various observation noise models $\mathbf{R}[k, \mathbf{z}(k)]$ is illustrated. The results of Figure 6.3 show how increasing the observation noise model stabilizes the location estimates for the stationary platform. The results of Figure 6.4 illustrate estimates for motion in the x-axis and y-axis and for motion which includes rotation. Estimates for motion

6.2 Estimation of Location

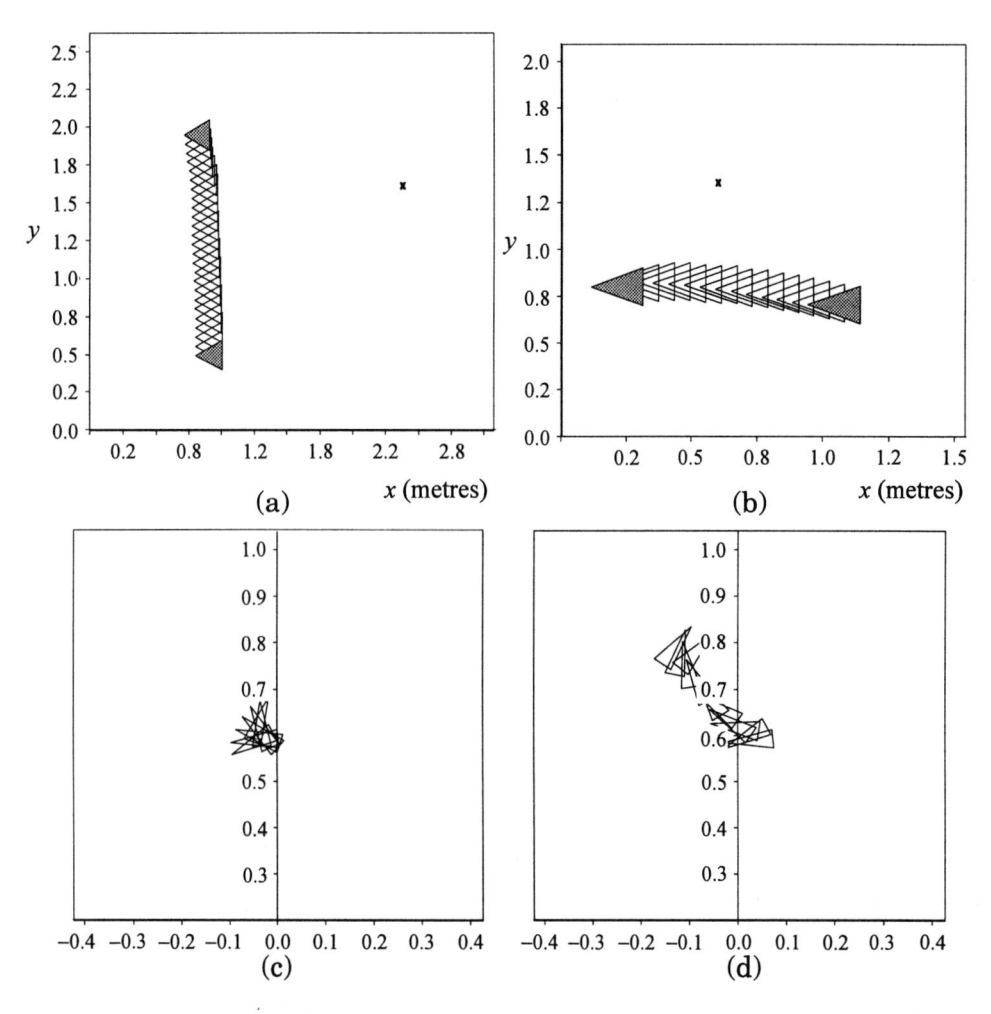

Figure 6.4: Localization estimates for (a) motion in y-axis and (b) motion in x-axis. In (c) and (d) localization estimates for sensor platform incorporating rotational motion i.e. (c) pure rotation (d) linear and rotational motion.

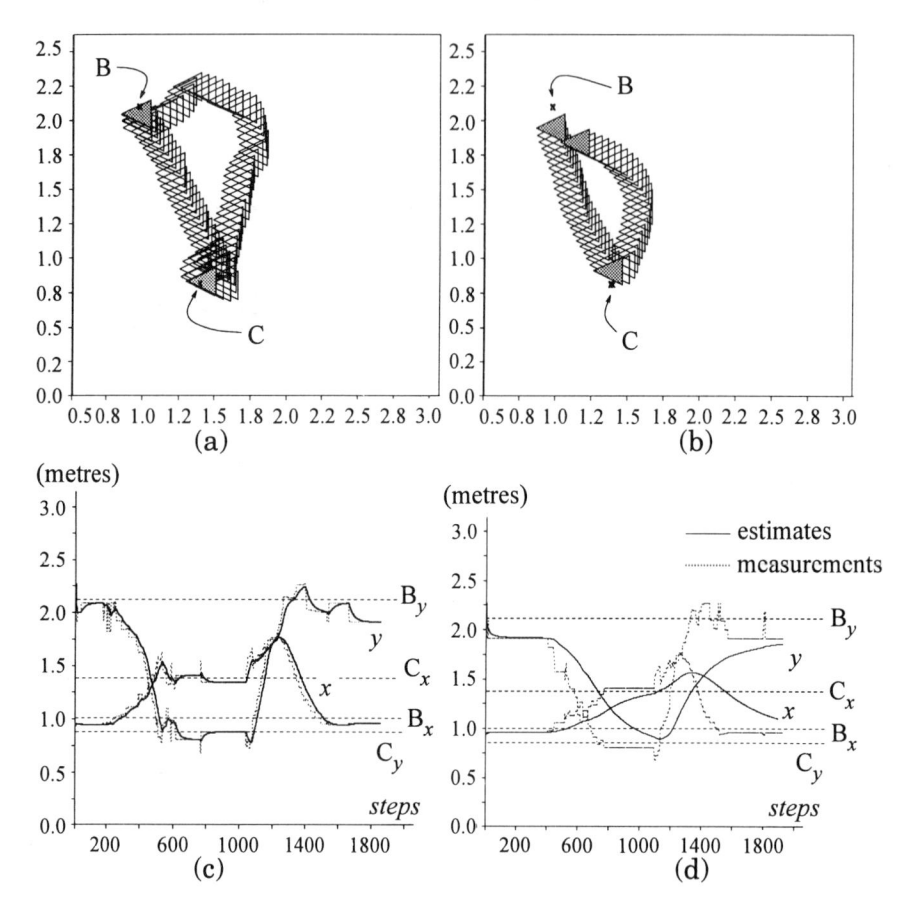

Figure 6.5: Location estimates during motion between two known locations i.e. from B to C and back to B stopping at C and B for a fixed period (10 seconds). The true (hand-measured) positions B and C are indicated. (a) and (b) show the estimated location of the platform. (c) and (d) show x y positions calculated directly from observed r and θ and the estimated x and y. The observation noise model $\mathbf{R}(k)$ is varied as follows; in (a) and (c) $\sigma_r^2 = 0.0001$, $\sigma_\theta^2 = 0.001$, in (b) and (d) $\sigma_r^2 = 0.05$, $\sigma_\theta^2 = 0.5$.

which includes pure rotations, linear and rotational motion are possible provided the control input $\mathbf{u}(k)$ of Equation 6.11 is given. The control input is obtained from kinematic control algorithms on the vehicle.

In order to evaluate the positional accuracy of the localization, estimates can be compared with hand measured locations. In the results shown in Figure 6.5, the platform moves from a known position B to a known position C stops momentarily and then moves back to B. The path followed between the points B and C is arbitrary.

The effect of the observation noise model on the accuracy of localization and speed of response can be seen in the results of Figure 6.5. Smaller values in the observation noise model give a faster response to actual changes in location but such a model is susceptible to biases and uncertainty in the actual observations. In practice therefore, the actual relative level of values chosen in the observation noise model depend on such factors as the motion and speed of the vehicle and the positional accuracy with which the feature used for localization is matched. It must be noted that if motion is unknown, i.e. $\mathbf{u}(k)$ is not known, the motion can be modelled by a relatively large process noise model [137]. This is a technique similar to use of the maneuver model [25][16] in which the kinematic model is stochastic. However, with a knowledge of the control $\mathbf{u}(k)$, better estimates can be obtained and motion which includes rotations can be included.

6.3 A Decentralized Localization Scheme

The localization algorithm of Section 6.2 can be implemented decentrally by making use of several Tracking Sonars mounted on the vehicle. The estimates from each sensor are fused using a decentralised information filter to give a global estimate of the position of the vehicle on which they are mounted.

6.3.1 Algorithm and Implementation

Each sensor operates as described in Section 6.2 by scanning and extracting RCDs. While the criteria discussed above for selecting the feature to be tracked still apply, other considerations such as sensor allocation need to be taken into account. Sensor allocation and sensor management for this application is discussed in Chapter 7. For the

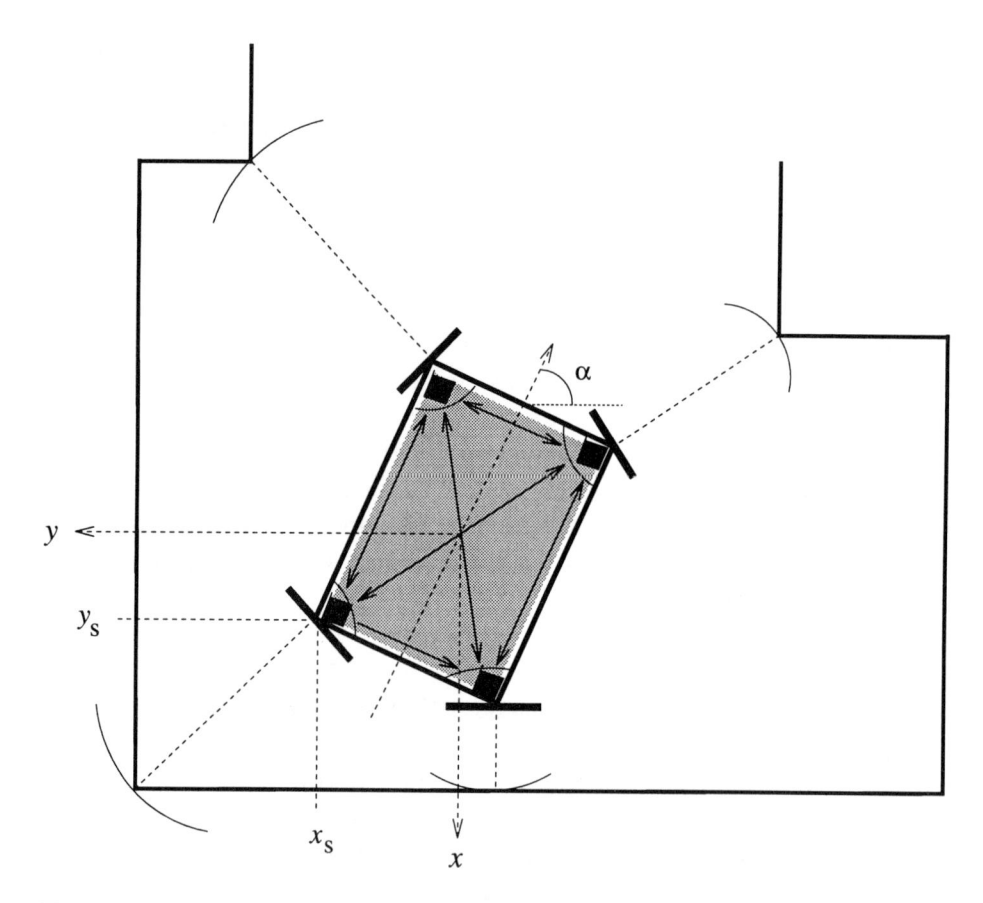

Figure 6.6: Tracking Sonar configuration for the OxNav vehicle. Each sensor focusses attention on a particular feature of the environment for purposes of localization. The sensors are decentralized and fully connected.

6.3 A Decentralized Localization Scheme

purposes of the discussion here, the sensor chooses to track the best matched RCD.

Each sensor currently tracking a matched RCD generates estimates of its *own* location rather than the location of the vehicle centroid. This necessitates the redefinition of the state to be the location of the centroid of the vehicle rather than that of an individual sensor (as in Section 6.2). A simple geometric transformation Φ is applied to the location of the sensor, i.e. $\mathbf{x}_s = [x_s, y_s, \alpha_s]^T$, to yield the location of the vehicle, provided that the location of the sensor on the vehicle is known. Such a transformation is of the form

$$\mathbf{x}(k) = \Phi \mathbf{x}_s(k). \tag{6.16}$$

Using such a transformation and the decentralized information filter (Equations 3.93-3.96), the state corresponding to the location and orientation of the vehicle can be estimated. For such a scheme, the sensors are configured as shown in Figure 6.6 for the OxNav vehicle (see Appendix B.4). Such a configuration is sometimes called single-platform-multi-sensor (SPMS).

Partial estimates are communicated only between the sensors that are *currently* tracking matched features. At each of the sensors, an estimate of the vehicle location $\hat{\mathbf{x}}_i(k \mid k)$ is obtained. This estimate is guaranteed to be the same at every sensor in the fully connected decentralized system. Several situations require consideration:

- Firstly, a sensor may extract RCDs which fail to match the expected ones. In this case the sensor may track the "most significant" RCD observed for purposes of map-building, i.e., track initiation and such a sensor is left out of the decentralized localization fusion.

- Secondly, a sensor may "lose" the RCD which it is currently tracking, as will occasionally happen, as discussed in Chapter 5. Such a sensor re-initializes by scanning and extracting RCDs and then attempting matches as before.

6.3.2 Decentralized Results and Performance

In all the results that follow, the estimates shown are obtained at a data rate which, for the 2-sensor system in these results, is approximately 25Hz. Results from a 2-sensor system are shown for ease of illustration.

Shown in Figure 6.7 is the effect of having two sources of observation, firstly, from different features being observed by two sensors and, secondly, from the same feature being observed by two sensors. In Figure 6.7 (b), the loss of the feature by sensor 2 can be explained by the interference which can result when a sensor receives acoustic signals from the other (see discussion in Chapter 5). Sensor 2 losing its feature has the effect that the estimates of the vehicle's location then tend towards those suggested by the remaining sensor's observations. The results of Figure 6.8 show estimates when the vehicle is moving linearly in the x-axis while localizing on two different features. In the results of Figures 6.7 and 6.8, it can be seen that the partial estimates $\tilde{x}_i$ are very close to the global estimates which themselves are identical as expected in a fully connected decentralized system.

The accuracy of the localization can be compared with hand-measured values in a sequence in which the vehicle moves to and from known locations using motion which is not smooth. The vehicle can be made to move in an "un-smooth" manner by pushing it along or deliberately introducing jerks in the kinematic control. Results from such un-smooth motion are useful as they test localization for robustness when vehicle motion is not properly modelled. The results in Figure 6.9 are for sensors localizing on the same feature at different ranges. It can be seen that at the closer range (Figure 6.9 (a) and (b)) the estimates are closer to the hand-measured values than at the longer range ((c) and (d)). The results of Figure 6.9 show that the accuracy decreases somewhat with range. More importantly however, in Figure 6.9 (c) and (d) it can be noted that although motion is only in the x-axis, at time $k = 900$, sensor 2 drifts off and starts tracking some other feature which results in a suggestion of motion, particularly noticeable in the y-axis, which is incorrect. These results demonstrate the unreliability and inaccuracy which may result through localization using the same feature at long range particularly when motion is not smooth. In practice, however, vehicles normally move in a smooth manner by being driven at constant speeds. In order to obtain localization results for such motion, the OxNav vehicle (JTR) is driven along a given trajectory. The results for such linear motion are shown in Figures 6.10 and 6.11 where it can be seen that the estimates are much smoother and closer to the measured ones.

To demonstrate further the accuracy and detail which can be achieved consider the following: When making turns on the OxNav

6.3 A Decentralized Localization Scheme

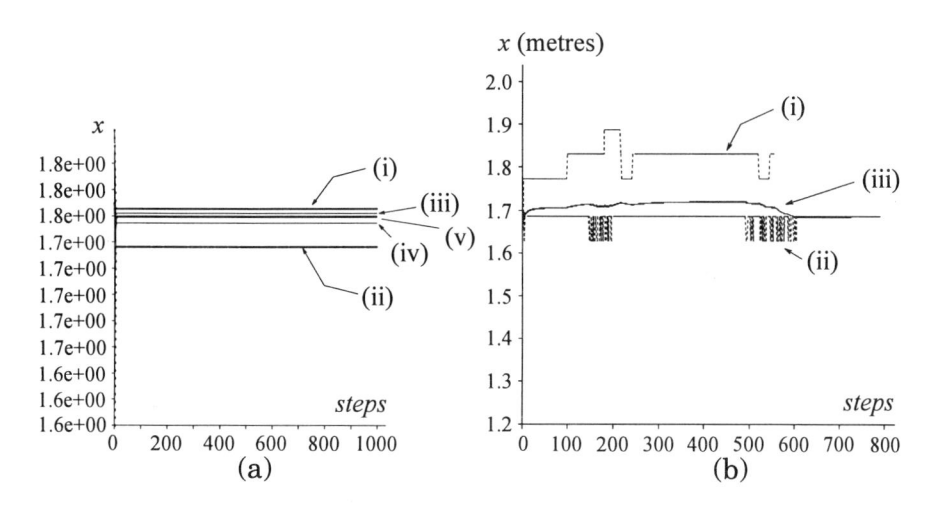

Figure 6.7: Decentralized location estimates (showing x-axis only) for a stationary vehicle using 2 Tracking Sonars. In (a) the sensors track different features and in (b) they track the same feature. In (b) sensor 2 loses the feature at time-step 550. The x-axis positions shown in (a) are; (i) a calculated directly from observations at sensor 1 and (ii) calculated from observations at sensor 2, (iii) and (iv) are the partial estimates at each sensor and (v) is the global estimate at each sensor. In (b) positions (iii) are the partial and global estimates.

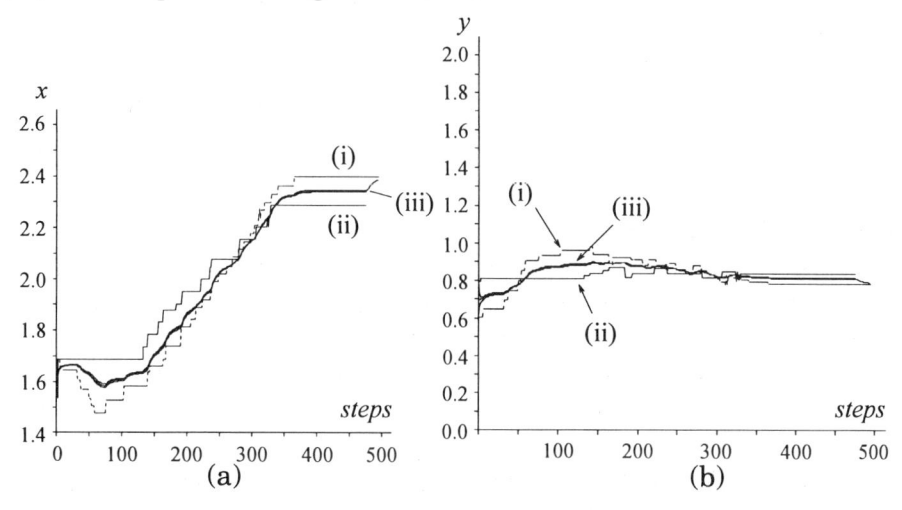

Figure 6.8: Decentralized location estimates for a vehicle which is moving in the x-axis. The x and y-axis positions shown are: (i) calculated directly from sensor 1 observations, (ii) from sensor 2 observations and (iii) are the estimates from the algorithm.

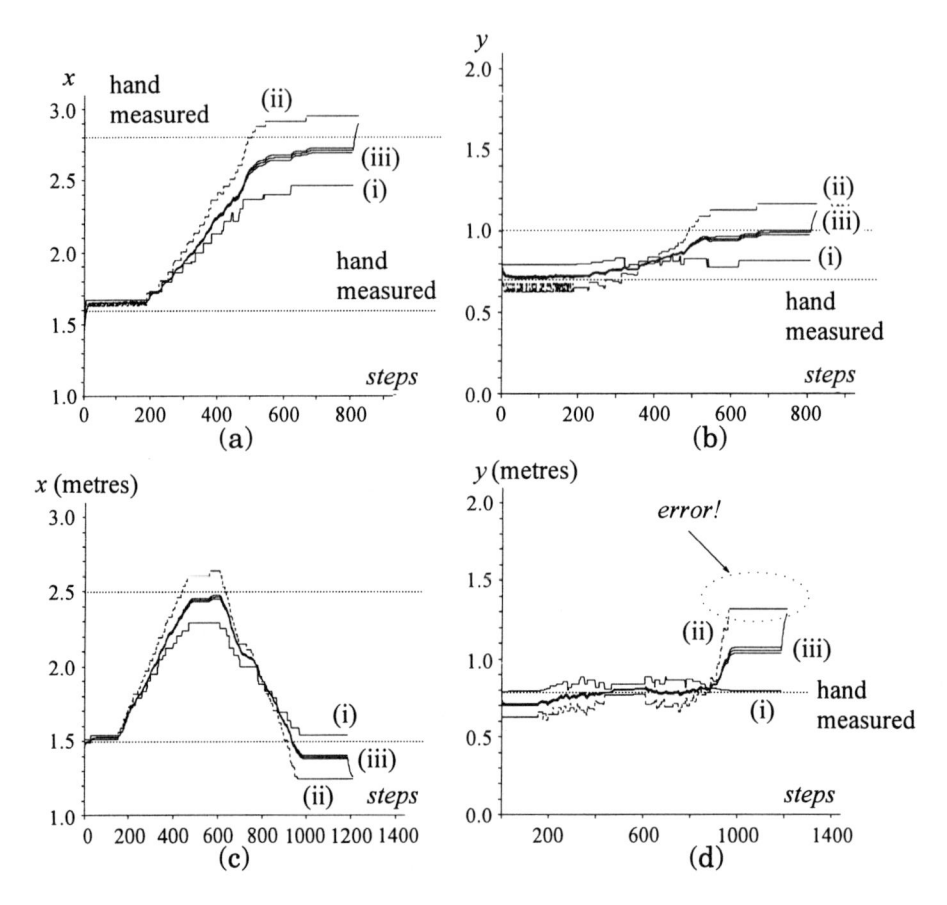

Figure 6.9: Comparing estimates with hand measured values while localizing on the same feature at different ranges. Here vehicle motion is not smooth. In (a) and (b) the feature being used for localization is at a range of 1.5m and motion is in both x-axis and y-axis. In (c) and (d), the feature is at a range of 4m and motion is only in the x-axis. In (c) and (d) sensor 2 at time $k = 900$, drifts and starts tracking some other feature causing an error in the estimate. The x and y-axis positions shown in are: (i) calculated directly from sensor 1 observations, (ii) from sensor 2 observations and (iii) are the estimates from the algorithm.

6.3 A Decentralized Localization Scheme

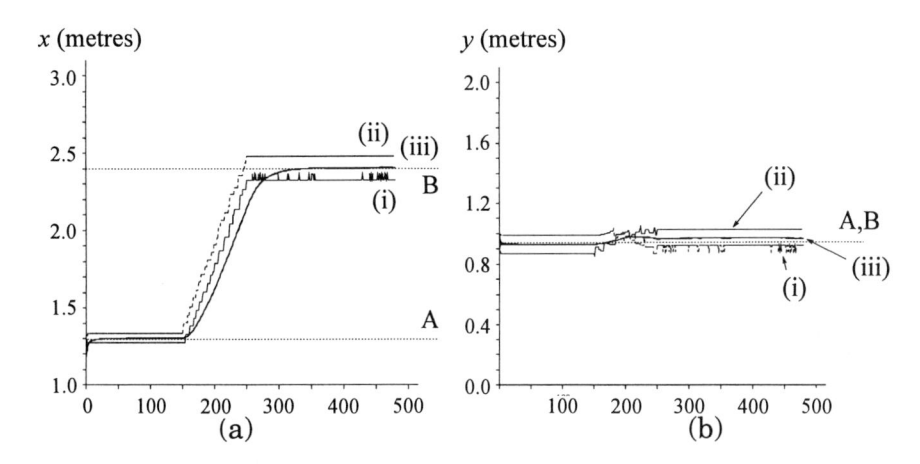

Figure 6.10: Decentralized location estimates with smooth vehicle motion while sensors track the same feature. Motion is in the x-axis only and (a) and (b) show the x and y estimates respectively. The x and y-axis positions shown are: (i) calculated directly from sensor 1 observations, (ii) from sensor 2 observations and (iii) are the estimates from the algorithm. Also shown are hand-measured values of positions reached.

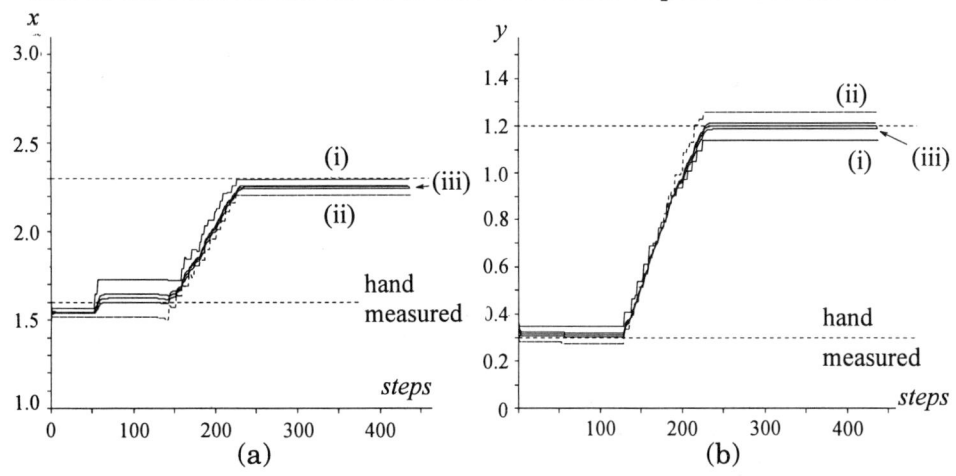

Figure 6.11: Decentralized location estimates with smooth vehicle motion while sensors track different features. Motion is in both x-axis and y-axis. Also shown are hand-measured values of the x and y positions reached.

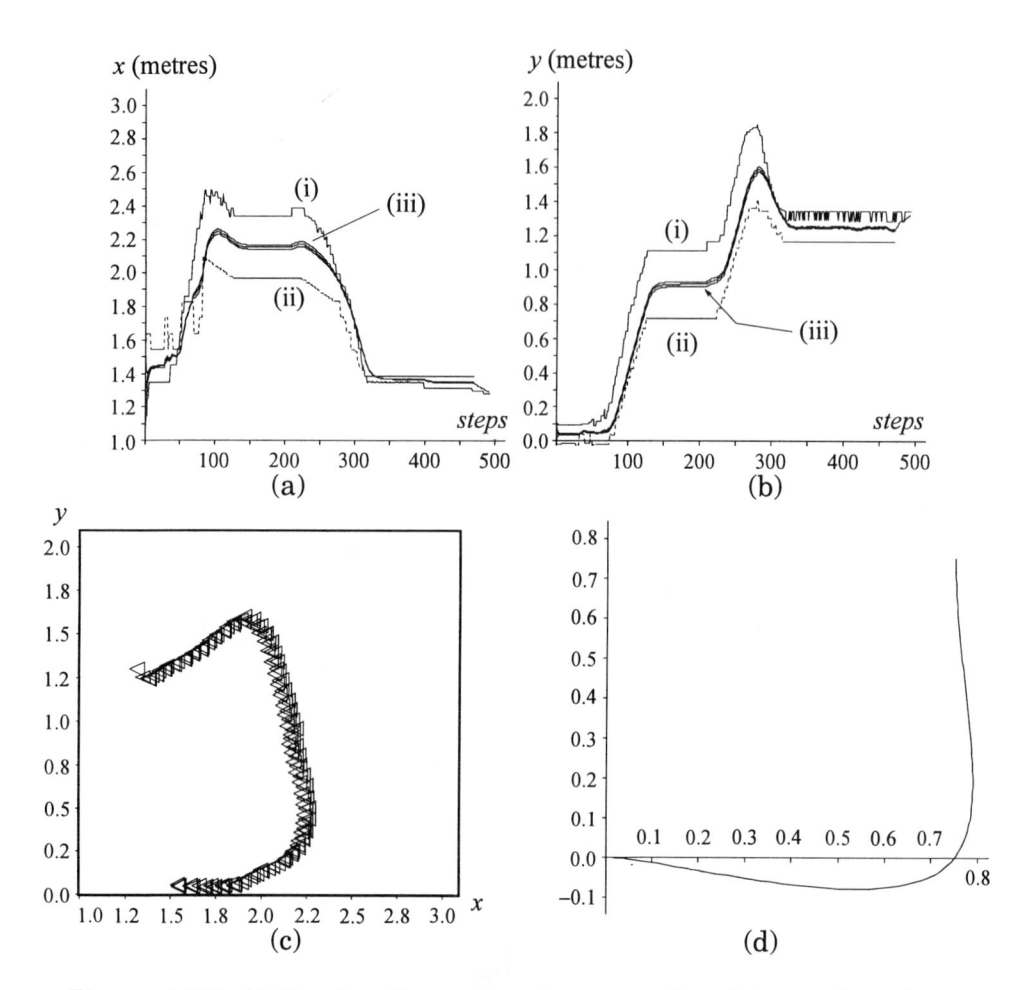

Figure 6.12: Estimates for motion incorporating 2 turns based on a trajectory as defined by a spline for turning a corner in a continuous motion. The x and y-axis positions shown are: (i) calculated directly from sensor 1 observations, (ii) from sensor 2 observations and (iii) are the estimates from the algorithm. The trajectory followed by the OxNav vehicle in executing the turns is defined by the spline shown in (d).

vehicle the trajectory can be made to follow a spline which is precisely defined. Therefore, the location of the vehicle can be estimated and compared with the trajectory defined by the spline. It can be seen in the results in Figure 6.12 that the estimates closely follow the path as defined by the spline which is plotted in Figure 6.12 (d).

Summary

In summary, the performance of the localization algorithm is dependent on

- *Nature of the vehicle motion and knowledge of* $\mathbf{u}(k)$. The estimation is enhanced when the kinematic control input $\mathbf{u}(k)$ in known and when the motion is smooth.

- *Allocation of sensors to features.* Better performance is obtained by ensuring that sensors do not estimate location based on concurrent observations of the same feature by multiple sensors.

- *Observation model.* The values used in the observation models which in turn are dependent on sensor operation parameters.

- *Continuous Validation.* This is validation which ensures that features used to estimate location at any given moment are the same ones matched. The concern here is that the focus of attention sensor, in this case the Tracking Sonar, may drift and track an adjacent feature.

The decentralized results presented demonstrate the following: Firstly, that the use of multi-sensors decentrally improves accuracy by providing estimates obtained by fusing observations from different viewpoints. Secondly, the decentralized algorithm performed better when sensors were tracking different features and this is reflected in the above results. Therefore, a good strategy is to ensure that sensors always localize on different features chosen for their informativeness.

6.4 OxNav Vehicle Navigation: A Discussion

Navigation, as described in this chapter, is being fully implemented and refined on the OxNav vehicle at Oxford[4]. Much of the software

---

[4]The OxNav group responsible for this vehicle and the implementation consists of T. Burke, H. Durrant-Whyte, J. Manyika, M. Stevens, and I. Treherne.

implementation work is due to M. Stevens. Figure 6.13 illustrates the vehicle.

Figure 6.13: The OxNav Vehicle *Joey*.

Mention has been made of the requirement for a facility to match extracted RCDs with those predicted from an *a priori* map. In a standard EKF-based matching algorithm, matching rates are far slower than the data rates attainable using the Tracking Sonars. However, given the scheme outlined in this chapter, this handicap is only encountered at initialization, i.e. when the sensor scans and extracts RCDs. Subsequently, matching is only required when there is a need to re-initialize. In addition, the rapid data streams obtained once tracking starts need to be validated. This is to ensure that the feature being being tracked is still the same one which was matched. This matching and validation requires rapid prediction of expected measurements, which in turn re-

6.4 OxNav Vehicle Navigation: A Discussion

scale: ⊢——⊣ 1 metre.

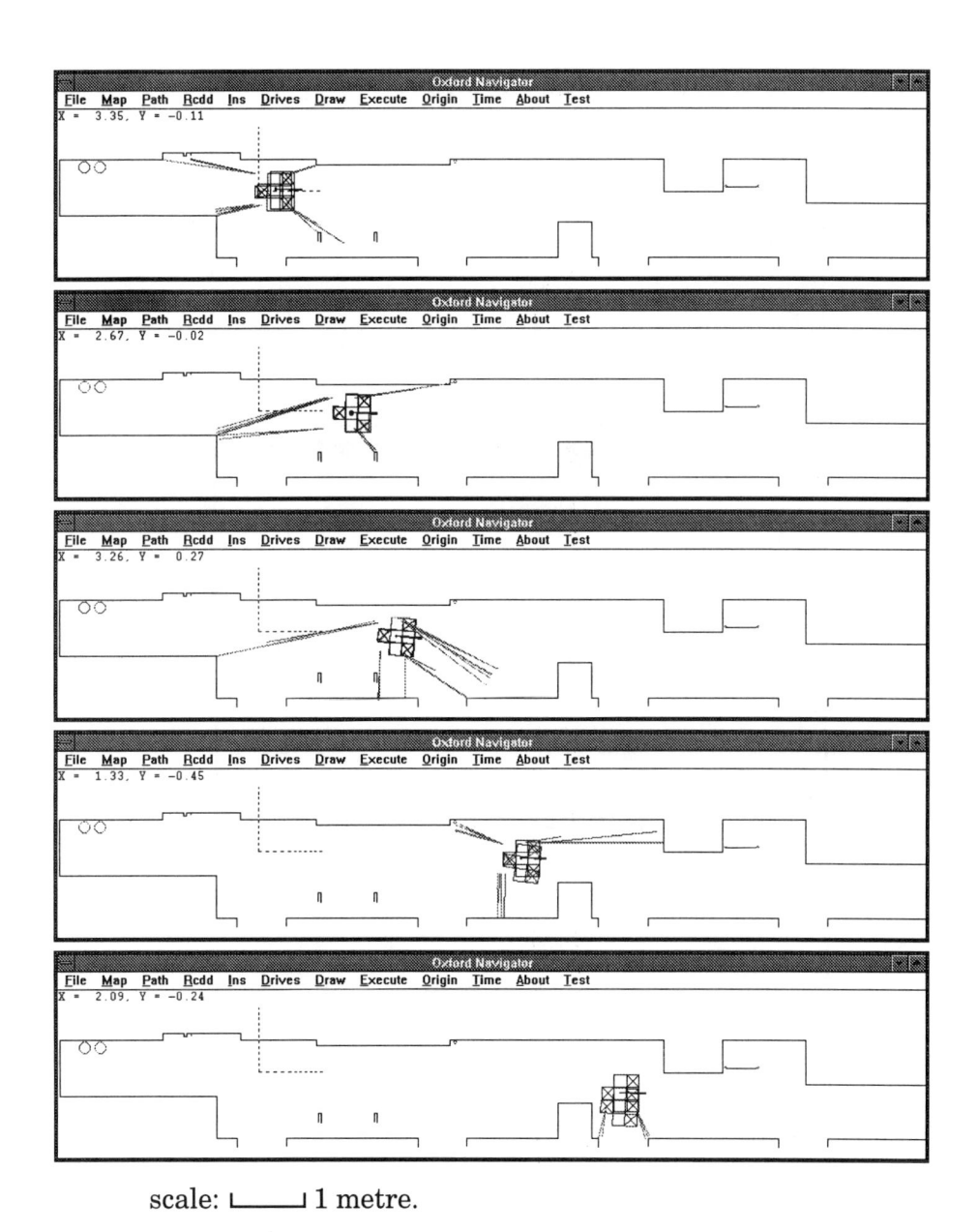

Figure 6.14: Joey navigating down a corridor. Snapshots from a navigation run down a corridor.

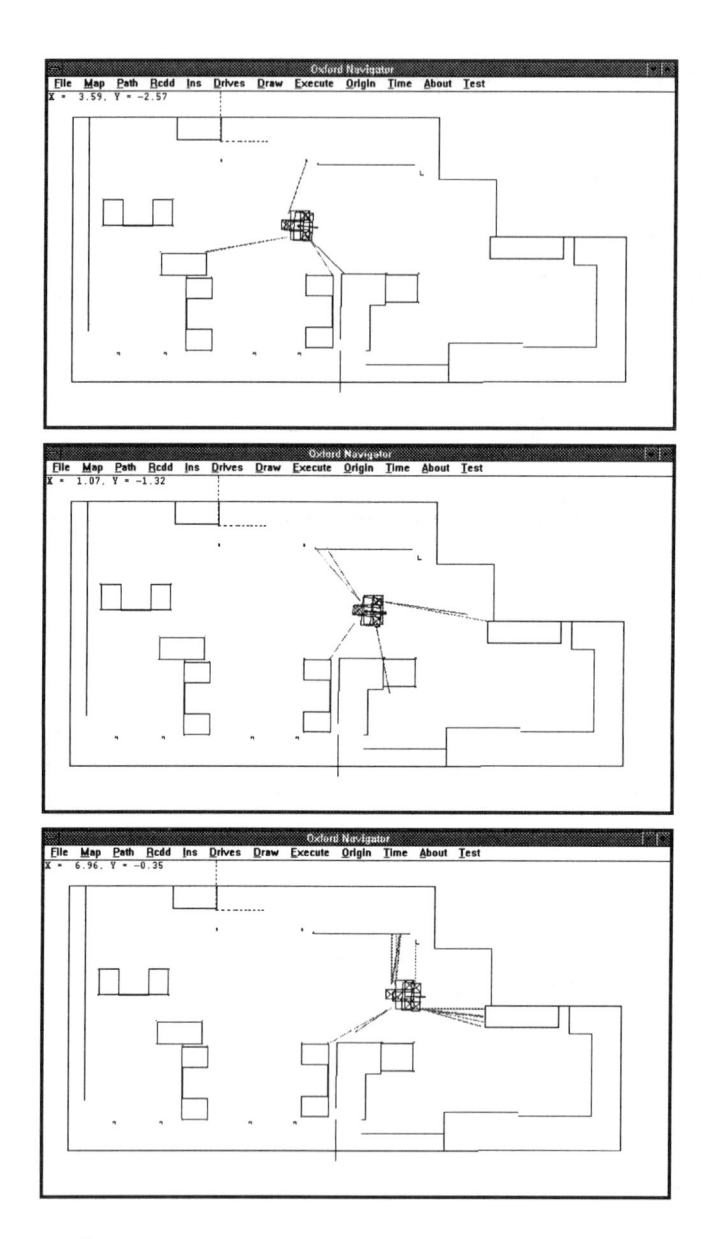

scale: ⊢——⊣ 1 metre.

Figure 6.15: Joey navigating in a cluttered environment.

6.4 OxNav Vehicle Navigation: A Discussion

quires a map representation space that can be searched quickly. This can be achieved in several ways;

- *Sensor geometric pruning.* Here sensor to feature and, conversely, feature to sensor visibility are used as criteria for eliminating features from the match or validation search, thus effectively pruning the search space.

- *Locality map segmentation.* This is appropriate in a large environment with a lot of features, whereby the search space can be reduced by considering only a subset of the map corresponding to the immediate locale.

- *Sensor-based map distribution.* Here map segments, as determined by FOV and position of sensor on the vehicle, can be distributed to each sensor and maintained an updated locally. (See discussion in *A Navigation Pandora's Box*, page 199.)

- *Numerical optimizations.* This involves the pre-computing of target invariants at the time that the map is generated. However, implicit in this is the assumption of a static environment.

All but the third one of the above have been implemented on the OxNav vehicle.

The OxNav vehicle has been tested in a variety of indoor environments. The environments have consisted of a variety of static features of different material. Figure 6.14 shows the vehicle navigating down a corridor. Figure 6.15 shows the vehicle navigating in a cluttered room. The figures show the estimates of position from the algorithm presented here and positions obtained from odometry. These estimates are consistent and nearly coincide with each other. In both cases the vehicle had an accurate map consisting of some of the features in the environment. In other results it was observed that, while estimates of location obtained using the decentralized sensors remained accurate, the location as given by odometry tended to become erroneous due to slip etc.

6.5 Modular Feature Classification

Focus of attention can be used to initiate tracks of previously unknown or un-associated RCDs in order to build maps. Thus, in addition to using the Tracking Sonar for map-building by tracking new features, map-building can also be enhanced by classifying the features being observed into their geometric types. In this respect, a modular algorithm for classifying features using the Bayesian Classification algorithm of Chapter 3 is now described.

6.5.1 Using Displacement to Classify Features

The fundamental idea in the algorithm is the exploitation of the displacement of the sensor due to motion to differentiate between types of features being observed. With observations from two displaced points, an algorithm can be developed to differentiate between *lines* and *points* using an approach based on triangulation. In our environment feature model consisting of planes, corners, edges and cylinders, observation of a line suggests that the corresponding feature is a plane and observation of a point suggests an edge, corner and even a cylinder with a small enough radius. This illustrated geometrically in Figure 6.16.

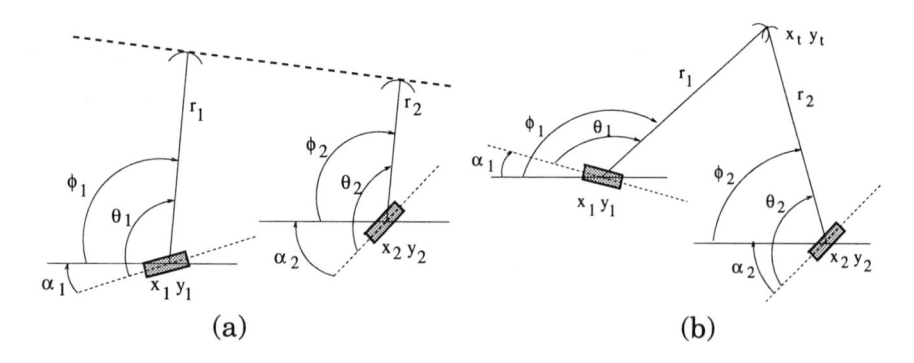

Figure 6.16: A classification model for observations from two positions (x_1, y_1) and (x_2, y_2). (a) shows the observation of a line from two positions and (b) observation of a point.

If a sensor observation is of a line, then the observation is as shown in Figure 6.16 (a). In Figure 6.16 (a), it can be seen that if $|\phi_1 - \phi_2| < \epsilon$, where ϵ is small enough then the observed feature is likely to be a plane. If the observation is as shown in Figure 6.16 (b), and the point

6.5 Modular Feature Classification

(x_{t1}, y_{t1}) coincides with (x_{t2}, y_{t2}) then the observation is likely to be that of a point. This provides a qualitative way of differentiating between lines and points. In order to develop a quantitative algorithm, consider Figure 6.17. It can be appreciated from Figure 6.17 that if the ratio $(\Delta_d/\Delta_p) \to 0$, then the observation is likely to be that of a line. If the ratio $(\Delta_d/\Delta_p) \to 1$, then the observation is likely to be that of a point[5]. This suggests using some thresholding technique in order to make the decision.

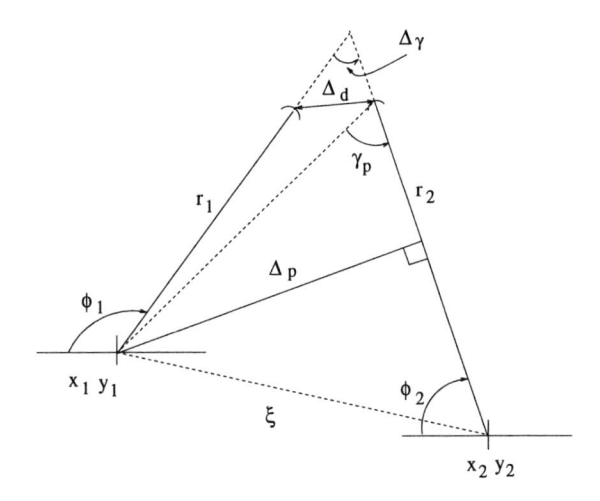

Figure 6.17: Observation model for an arbitrary feature illustrating the parameters used for classification. Observations are made from two positions (x_1, y_1) and (x_2, y_2).

However, one can do better by developing "variance" terms as follows

$$\sigma_d^2 \triangleq \left(\frac{\Delta_d - \Delta_p}{\Delta_p}\right)^2 \quad \text{and} \quad \sigma_\gamma^2 \triangleq \left(\frac{\Delta_\gamma - \Delta\gamma_p}{\Delta\gamma_p}\right)^2. \tag{6.17}$$

In this way, the variance σ_d^2 has values in the interval $[0, 1]$ with values close to 1 suggesting a point and values close to 0 suggesting a line. Similarly, σ_γ^2 has values in the interval $[0, 1]$, with values close to 1 suggesting a line and values close to 0 suggesting a point. Therefore, the likelihood of the observation being a line is given by

$$p(\mathbf{z}' \mid line) = \sigma_\gamma^2, \tag{6.18}$$

---

[5]The Δ variables used here are not to be confused with the differential measurement Δ, from Chapter 5.

and the likelihood of the observation being a point is given by

$$p(\mathbf{z}' \mid point) = \sigma_d^2, \qquad (6.19)$$

where $\mathbf{z}'$ is the set of observations, that is, $\{(r_1,\theta_1),(r_2,\theta_2)\}$, obtained from the two observation positions (x_1,y_1) and (x_2,y_2). In order to make these into probability distributions, they need to be normalized. In keeping with the notation of the algorithm in Chapter 3, the observation parameter model $\{M\}$ is given by $\{M\} = \{M_1, M_2\}$ where the parameters are $M_1 =$ line and $M_2 =$ point, and so the normalized likelihoods are

$$p(\mathbf{z}'(k) \mid M_1) = \frac{\sigma_\gamma^2}{\sigma_\gamma^2 + \sigma_d^2}, \qquad (6.20)$$

and

$$p(\mathbf{z}'(k) \mid M_2) = \frac{\sigma_d^2}{\sigma_\gamma^2 + \sigma_d^2}. \qquad (6.21)$$

The model relating these parameters to the state vector is $p(\{M\} \mid \mathbf{x})$, where the state is given by

$$\mathbf{x} = [X_1 = plane, X_2 = point]. \qquad (6.22)$$

The distinct classification $X_2 = point$, represents features such as corners, edges and cylinders whose radius is small compared to their range from the sensor, and so

$$point = \{corner, edge, small\ cylinder\}.$$

It is not possible to distinguish between elements of *point* from the above model. However, having decided that the observation is of a point, differentiation between edges, corners and cylinders can be achieved using visibility angles or matching with environment maps.

6.5.2 Algorithm and Implementation

The algorithm requires sensor displacement data. Such information can be obtained from various sources depending on the application. In the implementation discussed here, displacement data is obtained from the location estimates of the sensor making the observations.

6.5 Modular Feature Classification

Observation

The sensor makes an observation $\mathbf{z}(k) = [r_1, \theta_1]^T$ and the classification algorithm obtains the location vector $\mathbf{x}(k) = [x_1, y_1, \alpha_1]^T$ from the localization algorithm (see Section 6.2). We define a minimum displacement ξ which gives the minimum base-line used in Figure 6.17. Using the geometry of Figure 6.17, the following parameters are computed; $\{\Delta_d, \Delta_p, \Delta_\gamma, \Delta_{\gamma_p}\}$. And using Equation 6.17, σ_d^2 and σ_γ^2 are computed.

Likelihood Computation

From Equations 6.20 and 6.21, the likelihood vector $\Lambda_{\mathbf{z}(k)}$, is computed as

$$\begin{aligned} \Lambda_{\mathbf{z}(k)} &= p(\mathbf{z}(k) \mid \mathbf{x}) \\ &= [p(\mathbf{z}(k) \mid \mathbf{M_1}), p(\mathbf{z}(k) \mid \mathbf{M_2})] \begin{bmatrix} p(M_1 \mid X_1) & p(M_1 \mid X_2) \\ p(M_2 \mid X_1) & p(M_2 \mid X_2) \end{bmatrix}. \end{aligned}$$
(6.23)

Classification

The posterior distribution (or belief vector) is given by;

$$p(\mathbf{x} \mid \mathbf{Z}^k) = [p(X_1 \mid \mathbf{Z}^k), p(X_2 \mid \mathbf{Z}^k)], \qquad (6.24)$$

where for each X_b the posterior is computed as

$$p(X_b \mid \mathbf{Z}^k) = p(X_b \mid \mathbf{Z}^{(k-1)'}) \left[\alpha p(\mathbf{z}(k) \mid X_b)\right], \qquad (6.25)$$

where $(k-1)'$, refers to the previous time-step when the posterior was updated, as opposed to $(k-1)$ which refers to the previous time that the localization algorithm provided location estimates. The regularity with which the posterior is updated depends on ξ. As discussed in Chapter 3, the inferred classification is obtained from a MAP estimate i.e.

$$\hat{X} = \arg\max[p(X_1 \mid \mathbf{Z}^k), p(X_2 \mid \mathbf{Z}^k)]. \qquad (6.26)$$

In the implementation of the above algorithm, the step ξ determines the minimum acceptable displacement with which the algorithm can be used. If no motion is detected, then no updating occurs. One method of detecting motion is by testing $\mid x_1 - x_2 \mid < \xi$ and $\mid y_1 - y_2 \mid < \xi$. It is important when obtaining the displacement information from the localization algorithm of Section 6.2, to ensure that the value of $\xi \neq 0$. This is because often the estimate is unstable and so setting $\xi \approx 0$ would give a false impression of displacement where there has actually been no motion. Such a situation is illustrated in the localization results of Figure 6.3.

6.5.3 Some Classification Results

In the following results, the sensor platform moves while observing a feature which is to be classified as either a plane or a point feature.

For the first set of results, the displacement information is very accurate, having been estimated from linear and very smooth motion on the OxNav vehicle (see Figures 6.10 - 6.12) and actual measurement data as observations. In the results shown in Figure 6.18, the following model values were used for $p(\{M\} \mid \mathbf{x})$; in (a) and (c);

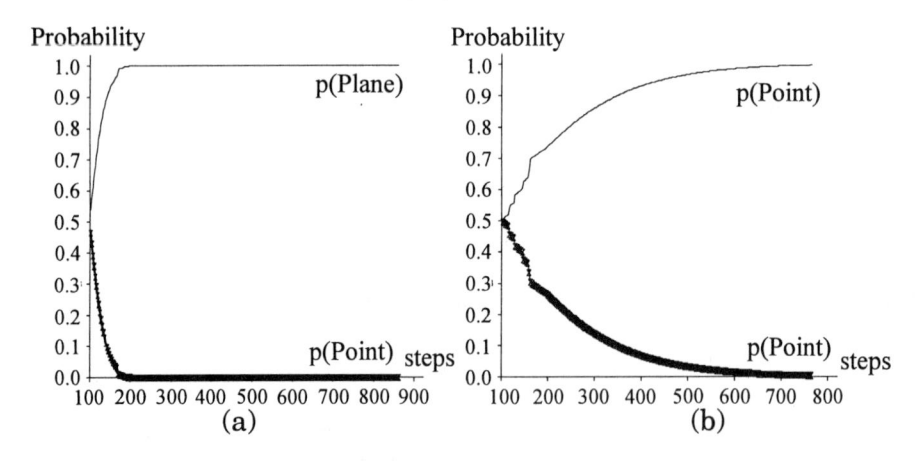

Figure 6.18: Variation with classification model for a plane. Probabilities for each feature type using very accurate displacement information (from very smooth linear motion). The models of $p(\{M\} \mid \mathbf{x})$ used are; in (a) model 1 and in (b) model 2.

$$\text{model 1}: \ p(\{M\} \mid \mathbf{x}) = \begin{bmatrix} p(M_1 \mid X_1 = line) & p(M_1 \mid X_2 = point) \\ p(M_2 \mid X_1 = line) & p(M_2 \mid X_2 = point) \end{bmatrix}$$

$$= \begin{bmatrix} 0.82 & 0.15 \\ 0.18 & 0.85 \end{bmatrix}, \tag{6.27}$$

6.5 Modular Feature Classification

and in (b) and (d)

$$\text{model 2}: \quad p(\{M\} \mid \mathbf{x}) = \begin{bmatrix} 0.52 & 0.45 \\ 0.48 & 0.55 \end{bmatrix}. \quad (6.28)$$

The probabilities are updated whenever motion with a minimum displacement ξ of 0.01m is detected. A displacement ξ of 0.01m is chosen in order to give an appreciable baseline. However, if the sonar data could be obtained precisely without any uncertainty and absolute displacement information could be obtained, then a smaller value of ξ could be used, a result which can be verified by simulation [146].

In Figures 6.19 and 6.20, the minimum displacement ξ was varied from 0.02m to 0.04m. In addition, the two models of $p(\{M\} \mid \mathbf{x})$ given in Equation 6.28 are used to illustrate their effect on classification using different values of ξ. The larger minimum displacement ξ of 0.04m results in correct classification for both the plane and the edge, using either model 1 or 2 for $p(\{M\} \mid \mathbf{x})$. Correct classification with ξ equal 0.02m is only obtained using model 1. When ξ is 0.02m is used together with model 2, the classification for the plane is uncertain as shown in Figure 6.19 (b) and the edge is classified incorrectly as a plane as shown in Figure 6.20 (b). This can be explained by the fact that the displacement data used in the algorithm derives from the localization algorithm and the degree of instability of this data is dependent on the models used in the localization algorithm and also in the uncertainty in the data observed as discussed in Chapter 5. These results highlight the sensitivity of the classification algorithm to the accuracy of the input displacement information. Similar results are obtained for other types of features.

Optimum performance of the classification algorithm can be obtained through judicious setting of the minimum displacement step ξ, taking into account the uncertainty in the displacement data. Knowledge of this uncertainty can be obtained from the observation model used in the localization algorithm which in turn depends on the Tracking Sonar parameters as discussed in Chapter 5. With the correct setting of these parameters, reliable classification can be obtained within the operating range of 0.40m to 4.00m with values of ξ greater than 0.03m. While setting the value of ξ very high provides a larger base-line, it is not desirable because then the algorithm is not updated frequently and so distinct classifications are only obtained after considerable periods. Therefore, the value of ξ in addition to localization performance,

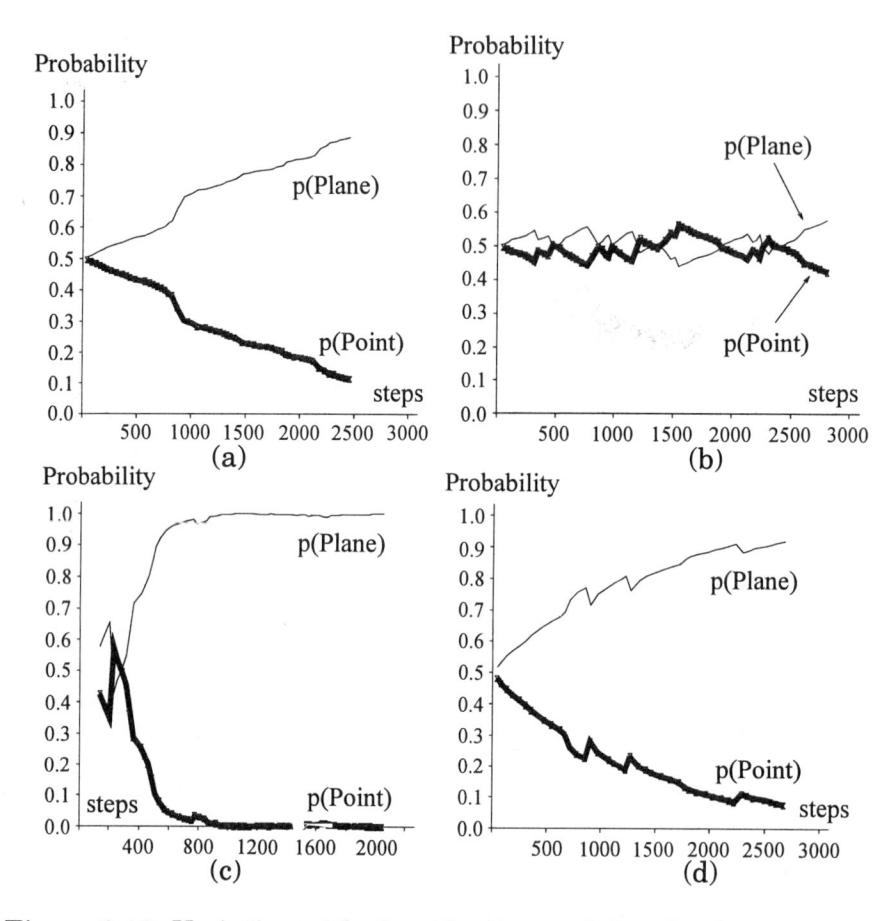

Figure 6.19: Variation with classification model and value of minimum displacement for a plane. Classifying a plane using two models of $p(\{M\} \mid \mathbf{x})$ and two values for the minimum displacement ξ. In (a) and (b), the minimum displacement $\xi = 0.02$m and in (c) and (d) $\xi = 0.04$m. In (a) and (c), model 1 is used for $p(\{M\} \mid \mathbf{x})$ and in (b) and (d) model 2 is used.

6.5 Modular Feature Classification

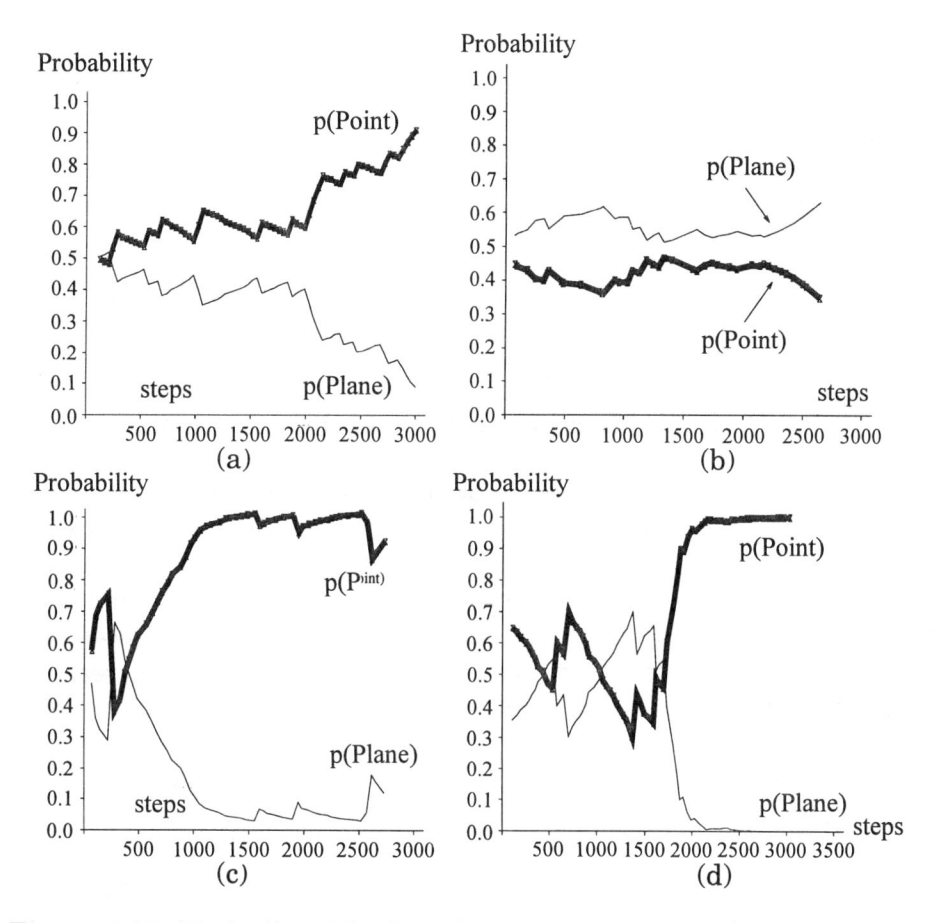

Figure 6.20: Variation with classification model and value of minimum displacement for a plane. Classifying an edge using two models of $p(\{M\} \mid \mathbf{x})$ and two values for the minimum displacement ξ. In (a) and (b), the minimum displacement $\xi = 0.02$m and in (c) and (d) $\xi = 0.04$m. In (a) and (c), model 1 is used for $p(\{M\} \mid \mathbf{x})$ and in (b) and (d) model 2 is used. In (b) the edge is incorrectly identified as a plane.

should also take into account the anticipated vehicle displacement and the urgency of the classification information in a given application.

Summary

The performance of the classification algorithm as implemented here is dependent on

- The value of the minimum displacement step ξ.

- Accuracy of the displacement data which in turn is dependent on the localization algorithm.

- The degree of certainty in the parameter model likelihood $p(\{M\} \mid \mathbf{x})$.

6.6 Concluding Remarks and Other Issues

The work presented in this chapter demonstrates the improvements to be gained from navigating using decentralized and distributed data fusion methods based on focus of attention, in this case Tracking Sonars. Notable, the previously significant overhead of maintaining map contact at all times by matching the observed features every processing cycle is eliminated. Hence estimation cycle time is greatly increased. The implementation of localization at the sensor level frees up the central system on a robot to perform other higher level functions such as path planning etc. The localization algorithms are augmented by the ability to classify features being tracked. This classification is limited to plane and point features and requires further development making use of information such as visibility angles to further refine classification. In a map-building scenario, the classification algorithm serves to augment the track initiation associated with incorporating new unknown features into a map. It is important to emphasize that the implementations presented here are not intended as robust and fully functional navigation systems, but as demonstrations of directed methods for navigation. Indeed navigation implemented as discussed here raises a host of other issues (see discussion in the box *A Navigation Pandora's Box* on page 199).

The algorithms that have been implemented in this chapter are not limited to vehicle navigation applications only. Indeed if the localization algorithm is reversed, i.e. if the sensors are stationary and the

6.6 Concluding Remarks and Other Issues

objects being observed are dynamic, then the system becomes like a typical multi-target tracking application. In such an application, the state vector is made up of sub-vectors each corresponding to each object that is being observed. This is the usual application of tracking algorithms normally encountered in the literature [16][25]. In fact such an implementation would be similar to the work presented by Rao [178] using CCD cameras and also discussed in [181].

Hopefully the application discussed in this chapter has demonstrated the need for sensor management which are the subject of the next chapter.

A Navigation Pandora's Box

Navigation implemented using the algorithms described in this chapter raises numerous interesting questions and avenues of exploration outside the scope of this monograph, not least of which are the following:

1. *How much decentralization?* The use of the focus of attention paradigm for localization does not imply the use of the decentralized information filter. The Tracking Sonars (or indeed any focus of attention sensor) can be used as simple sensors with fusion and estimation taking place at a central location. However by decentralizing the localization process, one could consider segmenting the environment map decentrally. This entails maintaining a local segment of a map at each sensor and updating it in a manner which is consistent with every other map segment at each sensor in the system. Several challenges and advantages arise from segmenting the environment map in this way.

2. *An Achilles heel.* The sacrifice made in pursuing a focus of attention exclusively, is that the robot then becomes completely unaware of anything else other than the features being tracked. Hence, new and potentially useful features as well as obstacles may be missed. Sensor management techniques can be employed in such a way as to make available a sensing capability for detecting new features and obstacles. Other possibilities include the use of a scanning sensor to augment a system using the focus of attention paradigm.

3. *Dynamic environments.* From the point of view of a Tracking Sonar (or any focus of attention sensor) the feature being observed can be stationary as well as in motion. What becomes a challenge is firstly, the problem of estimating location based on features whose motion or trajectories may not be known. Secondly, the question of updating and maintaining a map consisting of dynamic objects.

6.7 Bibliographical Note

This chapter has only outlined one application of the algorithms and architectures of Chapter 3, however numerous other applications are possible. The problems described in much of the target tracking literature are amenable to the methods and architectures that we have presented; reference can be made to the problems described in Bar-shalom [16, 14], Blackman [25, 26], Kenefic [122], Kurien [129] and Chong *et al* [49] and indeed the military applications surveyed by Waltz and Llinas [203]. In all these references, the multi-sensor methods are largely centralized or hierarchical with a only few, for example [49], describing distributed systems. For comparison, other approaches for addressing tracking problems are described in [128][157][190]. We have applied the architectures and algorithms presented here to other sensing problems and this work is described in Rao and Durrant-Whyte [179, 177] for a multi-camera system, Grime and Durrant-Whyte [91] and Gao and Durrant-Whyte [84] for a process plant, Burke and Durrant-Whyte [39] for robot kinematic control.

The literature on navigation is extensive and the approaches adopted vary considerably. Beck [21] gives an introduction to navigation systems. For descriptions of some of the methods used, reference can be made to the work by Anderson [7], Barshan [18], Chatila [46], Giralt *et al* [88], Thorpe *et al* [201] and Waxman [204]. The approach we have taken to robot localization is largely based on the work by Leonard and Durrant-Whyte [137] which we use as a starting point. Much of the original work on sonar based localization was done by Drumheller [66], Crowley [58, 57], Elfes [73] and Kuc [126]. Reference can also be made to other related work in localization done by Grimson and Lozano-Pérez [92]. We have not fully described map-building but only the process of classifying features; more discussion on map-building can be found in Brooks [34], Leonard [135] and Cox [56]. Other descriptions of feature extraction and classification can be found in [36] and [179].

7 Sensor Management Demonstrations

*A thing may look specious in theory,
and yet be ruinous in practice;
a thing may look evil in theory,
and yet be in practice excellent.*

- Edmund Blake

While distributed and decentralized sensor management is clearly desirable, its implementation has hardly been demonstrated in systems other than simulations. Having so far presented 'specious' methods for managing data fusion, this chapter aims to demonstrate the successful application of these methods to a practical sensing problem.

In demonstrating sensor management, it is important to reiterate the practical reasons for managing multi-sensor systems:

- **Sensor-Feature Assignments.** It is important to manage and coordinate the assignment of sensors to targets (or features) in order to use sensing resources effectively. This also ensures that all the features that need to be observed are covered in a manner which is consistent with system goals. To be most effective, the management of sensor-feature assignments must take into account the dynamic nature of the problem, such as in the case of moving targets or moving sensor platforms, by continually reviewing current assignments.

- **Sensor Cueing and Hand-off.** When using sensors with limited fields of view it is important to ensure that features or targets which may pass out of view are not lost. Hence, it may become necessary to *cue* sensors into whose field of view a feature may be entering. Cueing may be done in a cooperative manner e.g.

when sensors, capable of obtaining different information, cooperate to resolve ambiguity concerning a particular feature. *Hand-off* refers to the transfer of the observation of a feature by one sensor to another. As expected, sensor cueing and hand-off should be consistent with system goals.

- **Multi-mode, diversity management.** Often sensors are capable of operating in several modes, therefore, it becomes necessary to make decisions concerning the most appropriate mode for a given situation. This is similar to the situation where there are several physically diverse sensors available. In such cases it becomes necessary to manage such diversity i.e. make decisions regarding the appropriate sensor or sensor mode for a particular observation.

In all the above, the criteria for management determined by the system goals. In the approach presented in this monograph, the system goal is that of obtaining and refining information about the state of nature. Therefore, sensor-feature assignments, cueing and hand-off are considered so as to be consistent with gaining information about the state of nature.

The problem discussed in this chapter concerns the management of distributed sensors on a mobile robot, which is fusing sensor data for purposes of localization and map-building. The sensor configuration for this application is that illustrated in Figure 6.6. This particular problem is representative of multi-sensor applications in which sensor management can play a pivotal role. The fundamental elements here are; a multi-sensor system, algorithms implementing data fusion, and a set of possible sensing actions. For data fusion we employ the algorithms of Chapter 3 applied to localization and feature classification, as discussed in Chapter 6 while making use of the sensor developed in Chapter 5. And for sensor management we turn to the methods of Chapter 4.

7.1 Sensor Management on a Navigating Robot

It is generally acknowledged that demonstrating implemented sensor management activity is something of a challenge [172]. Demonstrating the advantages of having sensor management is made difficult by its feedback nature in the data fusion loop as illustrated in Figure 1.3.

Nevertheless, one method which can be employed is to show the quantities on which management decisions are based and also the decisions themselves, for a variety of situations. Such results must show that the actions adopted are consistent with the management imperative. In our context, this amounts to showing that the resulting management decisions always maximize information and reduce uncertainty with regards to the state of nature.

The results presented are based on the implementation of vehicle navigation using Tracking Sonars mounted on the OxNav vehicle (JTR) as described in the previous chapter. For localization, the information metrics on which management is based are with respect to the state $\mathbf{x} = [x, y, \alpha]^T$, i.e. the location of the vehicle. This state can be estimated decentrally using sensors which make observations of known features in the environment by focussing attention on such features. A typical environment for this indoor application may contain several (of the order of 10) features that can potentially be used to estimate location. Clearly, such a system equipped with a finite number of focus of attention sensors, say 4, can observe only a subset of the features available. The allocation and coordination of the sensors in this situation represents the management problem. We also present management results from the implementation of the feature classification algorithm described in Chapter 6. For this algorithm, the information metrics are with respect to the state $\mathbf{x}$, which is the classification of a particular feature being tracked while localizing. Therefore, given our discussion in Chapter 4, we can state from the outset that the management of these two algorithms cannot be coupled because the states refer to two distinct entities as described by **Case 2** in Section 4.5.3.

7.2 Quantities for Sensor Management

7.2.1 Variations in State Estimation Metrics

The information obtained by a sensor depends on its ability to take measurements and the uncertainty associated with these measurements. For the Tracking Sonar, the uncertainty is modelled by the covariance of the measurement noise given by $\mathbf{R}$. In Chapter 5, we showed that the variances in $\mathbf{R}$, depend on the sensor operational parameters $\zeta = (\sigma_\delta^2, \sigma_u^2, \psi)$. The choice of these parameters depends on the tracking performance required. For a fixed range and orientation, and known parameters ζ, the variances σ_r^2 and σ_θ^2 can be approximated based on the analysis of Chapter 5. In addition, these variances are scaled by factors which depend on the values of r and θ themselves. This scaling can be modelled by $g_r(r)$ and $g_\theta(\theta)$. Hence, the covariance, written as $\mathbf{R}[k, \mathbf{z}(k)]$, can be considered as a *location covariance*, whose trace gives the actual (scaled) variances of r and θ respectively. For a given set of operating parameters and a known operation range, the variances used in Equation 5.18, can be approximated.

The results that follow show how the information metrics $\mathbf{I}(k)$, $\mathbf{i}(k)$ vary under different conditions. It must be noted that, since entropy for continuous distributions is relative [52], the absolute values of the information measures $\mathcal{I}(\cdot) = -h(\cdot)$, are not significant and only their relative values are of importance.

Variation with range. Figure 7.1, shows how the information quantities vary with range and the effect that different location covariance models have on the metrics and consequently management decisions. In Figure 7.1 (b), it can be seen that the posterior information $\mathbf{I}(k)$, decreases slightly as the range is increased. This is because the information contained in each observation $\mathbf{i}(k)$ decreases with range. The cumulative observation information $\sum_k \mathbf{i}(k)$, if plotted, continues to grow but at a slightly reduced rate. The effect of the different models can be seen quite clearly. A model which reduces the variance of the measured quantities results in relatively higher quantities of posterior and observation information $\mathbf{I}(k)$ and $\mathbf{i}(k)$ respectively. Considering the cumulative information gain is useful because it shows trends in the way the information is gained. The usefulness of this shall be shown in later results.

Variation with field of view. Figure 7.2, shows how the information quantities vary across the field of view by moving the sensor across in front of a tracked feature as shown in (a). It can be seen in (b) that the information $\mathbf{I}(k)$ increases as the sensor moves towards the feature and starts to decrease as the sensor moves away from the feature. This is because the observation information $\mathbf{i}(k)$ is greatest when the sensor is closest to the tracked feature as shown in Figure 7.2 (c).

The results of Figures 7.1 and 7.2 show that the information obtained by each sensor depends on; (i) the observation model which in turn is related to the sensor operational parameters and (ii) the relative spatial

7.2 Quantities for Sensor Management

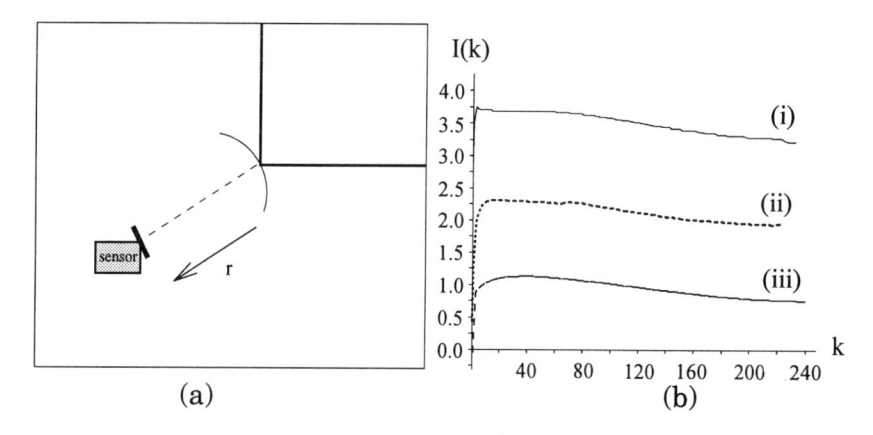

Figure 7.1: Variation of information quantities with range for various observation models. The motion of the vehicle is as shown in (a). The graphs show in (b) global information $\mathbf{I}(k)$ Values in $\mathbf{R}(k)$ are as follows; in (i) $\sigma_\theta^2 = \sigma_r^2 = 0.0001$ (ii) $\sigma_\theta^2 = \sigma_r^2 = 0.001$ and (iii) $\sigma_\theta^2 = 0.005$ $\sigma_r^2 = 0.01$.

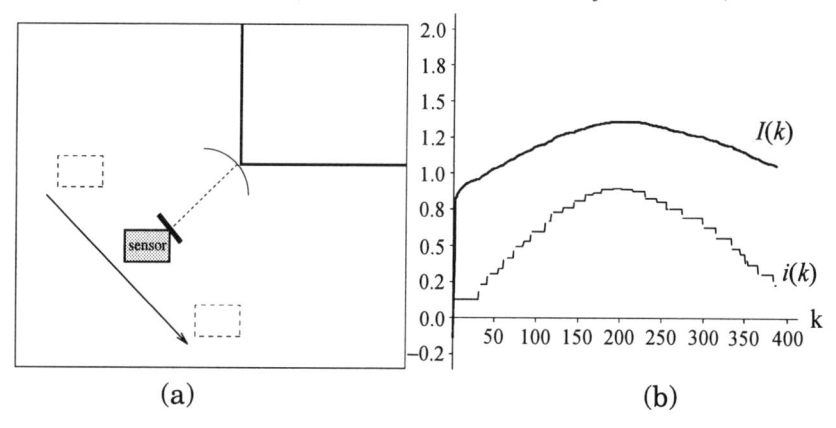

Figure 7.2: Information quantities for motion past a tracked feature. (a) shows the motion past the feature and (b) the global information $\mathbf{I}(k)$ and the observation information $\mathbf{i}(k)$ for $\sigma_\theta^2 = 0.005$ $\sigma_r^2 = 0.01$.

locations between the tracked feature and the sensor.

7.2.2 Variations in Classification Uncertainty Metrics

The classification algorithm implemented in Chapter 6 depends on the parameter model $p(\{M\} \mid \mathbf{x})$ and vehicle location information in the following way: Firstly, the parameter model $p(\{M\} \mid \mathbf{x})$ relates the parameters $\{M\}$ to the state and thus incorporates the uncertainty inherent in the geometrical interpretation of the observed parameters. Secondly, implemented as in Chapter 6, the classification algorithm makes use of location information from the localization algorithm, to measure displacement. A displacement step ξ is defined as the minimum displacement for the update of the classification algorithm. Since the location information is not precise, the results of Chapter 6 showed that a larger, rather than smaller, step is expedient because it limits the effect of the location estimation uncertainty. In the following results, we vary both these factors and examine their effect on the information metrics used for management.

Variation with parameter model. In the following results the same parameter models as in Chapter 6 are used, i.e.

$$\begin{aligned}
\text{model 1}: \; p(\{M\} \mid \mathbf{x}) &= \begin{bmatrix} p(M_1 \mid X = line) & p(M_1 \mid X = point) \\ p(M_2 \mid X = line) & p(M_2 \mid X = point) \end{bmatrix} \\
&= \begin{bmatrix} 0.82 & 0.15 \\ 0.18 & 0.85 \end{bmatrix} \\
\text{model 2}: \; p(\{M\} \mid \mathbf{x}) &= \begin{bmatrix} 0.52 & 0.45 \\ 0.48 & 0.55 \end{bmatrix} \\
\text{model 3}: \; p(\{M\} \mid \mathbf{x}) &= \begin{bmatrix} 0.65 & 0.31 \\ 0.35 & 0.69 \end{bmatrix},
\end{aligned} \tag{7.1}$$

where model 3 is added for illustrative purposes. In the results of Figure 7.3, we show the information quantities $\mathbf{I}(k)$ and $\mathbf{i}(k)$ obtained as the sensor moves linearly past the feature being classified in the same way as the motion in Figure 7.2 (a). The results show that with increasing confidence in the parameter model $p(\{M\} \mid \mathbf{x})$, the posterior entropy $(-\mathbf{I}(k))$ is decreased more rapidly. With model 1 (highest confidence), the initial observations contain a lot of information as shown

7.2 Quantities for Sensor Management

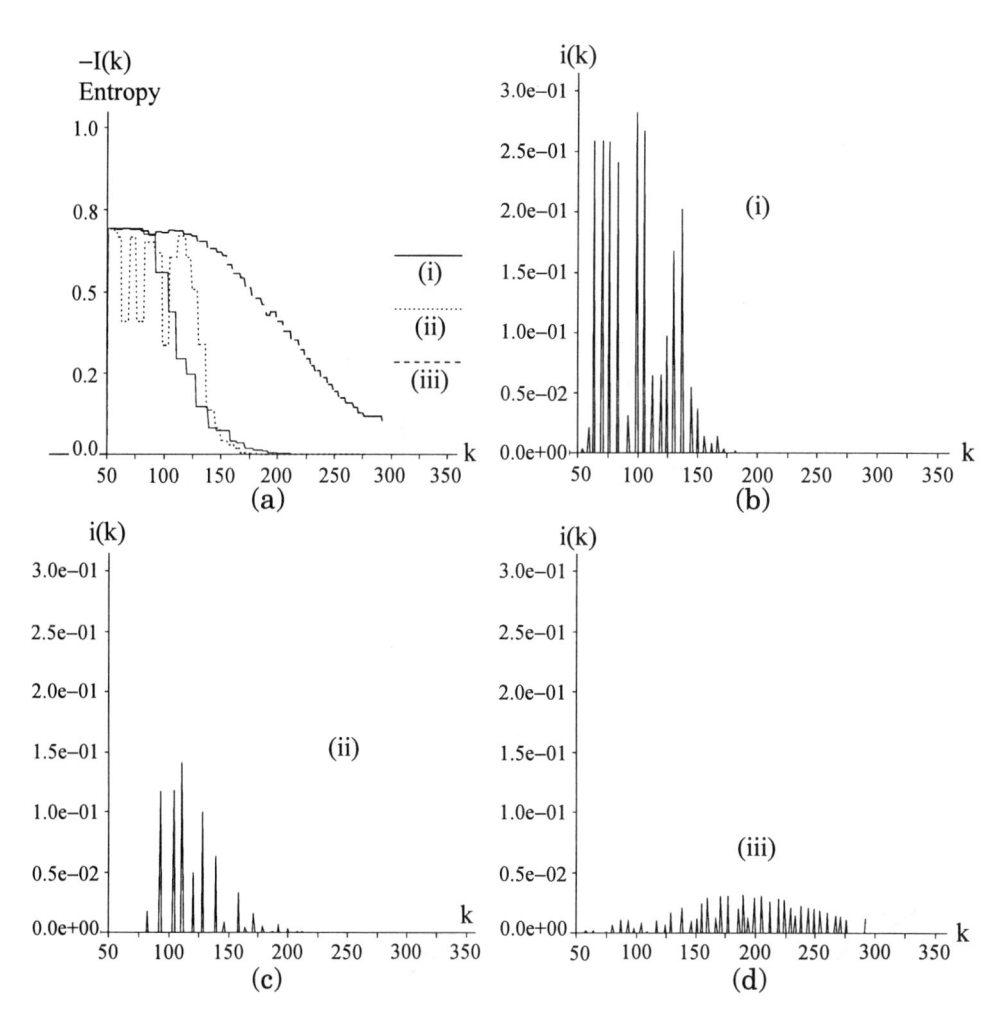

Figure 7.3: Variation of classification information with parameter model. Motion is linear past a tracked feature being identified. (a) Entropy (negative information), (b), (c) and (d) observation information. (i) is model 1, (ii) is model 3, and (iii) is model 2.

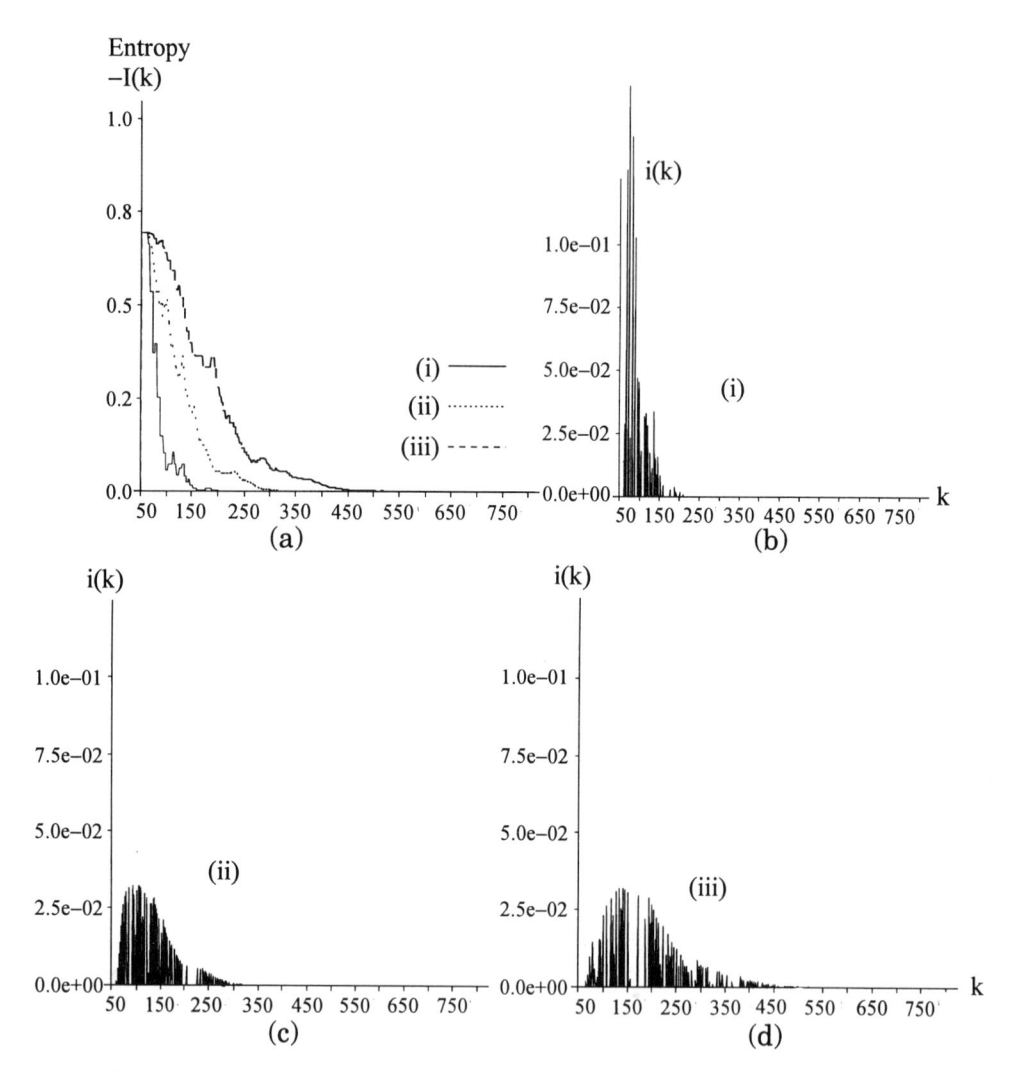

Figure 7.4: Variation of classification information with displacement step ξ. Motion is linear past a tracked feature being identified. (a) entropy (negative information), (b), (c) and (d) observation information. For (i) step = 0.03m, (ii) step = 0.02m and (iii) step = 0.01m.

7.2 Quantities for Sensor Management

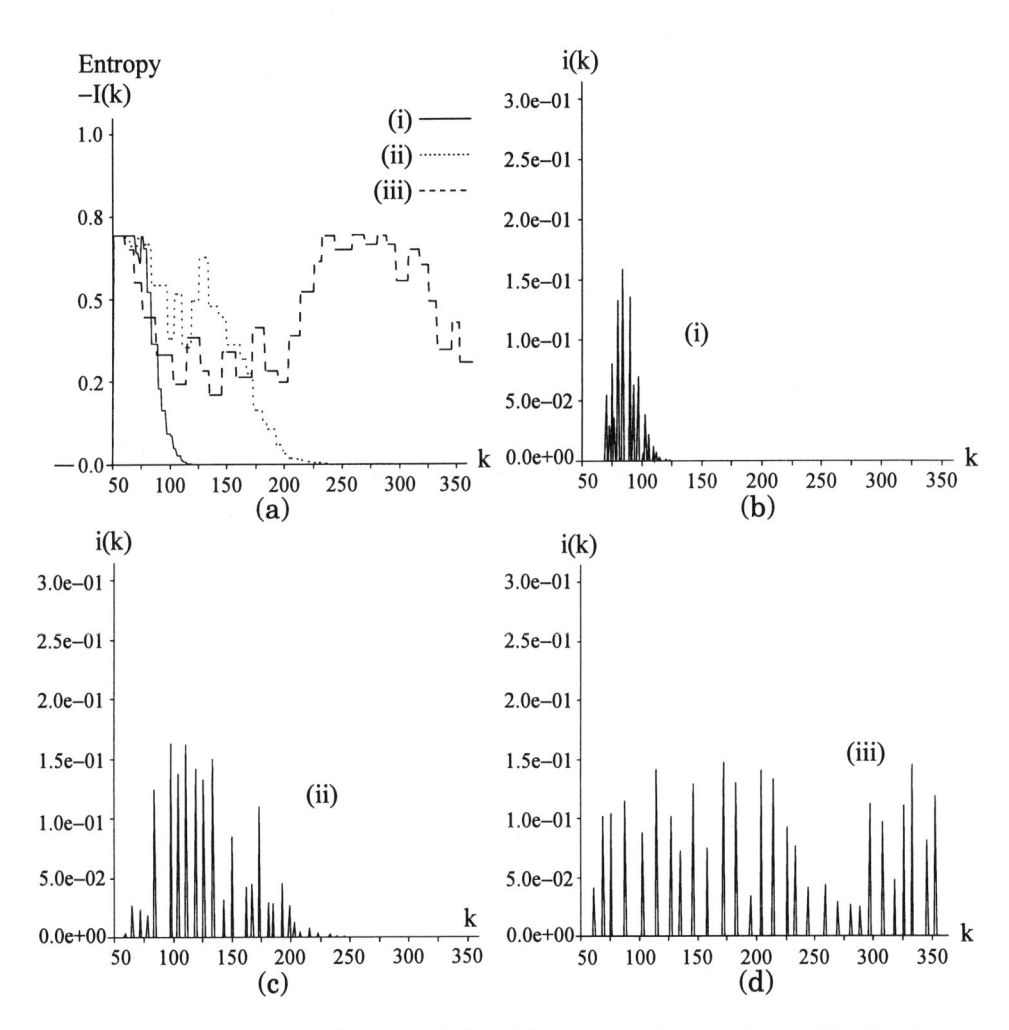

Figure 7.5: More variation of classification information with displacement step ξ. Motion is arbitrary in front of tracked feature. (a) Entropy (negative information) (b) (c) and (d) Observation information. (e) Observation information gain. For (i) step = 0.03m, (ii) step = 0.02m and (iii) step = 0.01m.

by $\mathbf{i}(k)$, and as uncertainty decreases, the informativeness of the observations decreases to negligible quantities at about $k = 175$ as shown in (b). Using model 2, useful, albeit low information content, observations continue to be made up to $k = 300$. This is also reflected in the information gain plots (Figure 7.6 (a)) where using model 2, information gain continues to rise.

Variation with displacement step ξ. In Figure 7.4, the minimum displacement step ξ, is varied for each run of a sensor platform moving linearly past the feature being classified. We make use of three typical values of ξ. It can be seen in (a) that entropy is more rapidly reduced with the larger value of ξ even though this means that updates occur less frequently compared to using a smaller step value. This is because feature classification, as implemented in Chapter 6, is better posed with a larger separation between the two observation positions used, as shown in Figure 6.17. The information in the observations reflect this as shown in Figure 7.4 (b), (c) and (d). In Figure 7.5, the same step values as above are used but the vehicle motion is arbitrary. With arbitrary motion, the location information is even less accurate, because sudden changes in direction or motion are not always reflected immediately in the location estimates (due to the estimation lags which vary with $\mathbf{R}$, see results in Chapter 6). With a step of 0.03m, the rate of uncertainty reduction is not greatly affected. With a step of 0.01m, entropy decreases and then increases due to ambiguities in the classification of the feature, this is also reflected in the observation information. An example of such ambiguities in classification was illustrated in the results shown in Figure 6.20 (b).

7.2.3 Discussion

The results presented illustrate the effect of various models and parameters on the information obtained and their effect in reducing uncertainty (entropy). The results also show the effect of relative spatial locations on the information obtained. This justifies the fact that if several such sensors are used, as in the decentralized implementations of Chapter 6, some sensors will be better suited than others for certain sensing tasks by virtue of their spatial locations and their ability to make observations.

7.3 Sensor-Feature Assignments

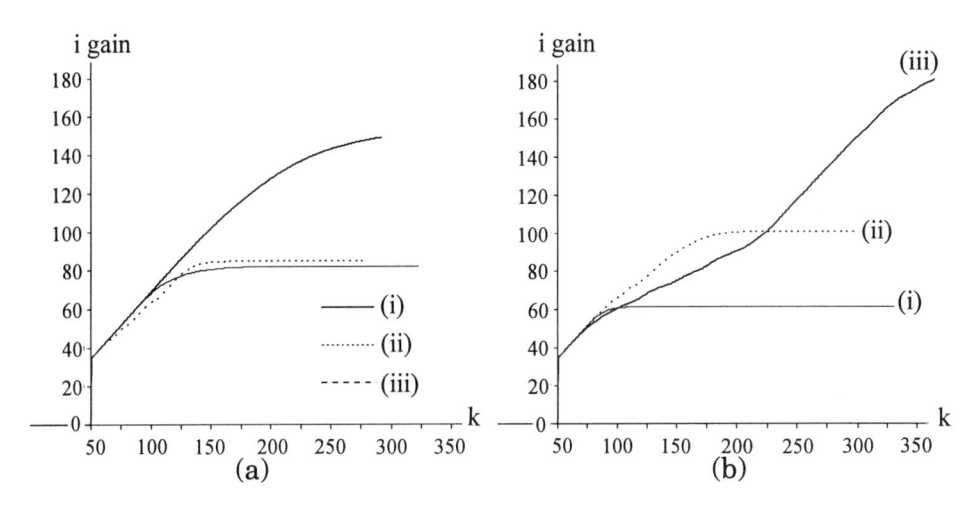

Figure 7.6: Cumulative observed information. (a) for variation with parameter model (corresponds to Figure 7.3). (i) is model 1, (ii) is model 3, and (iii) is model 2. (b) for variation with accuracy of location information (corresponds to Figure 7.5). In (i) step = 0.03m (ii) step = 0.02m and (iii) step = 0.01m.

7.3 Sensor-Feature Assignments

In the algorithms developed in Chapter 4, decisions regarding sensor-feature assignments are made by choosing assignments which maximize the information obtained. Here we demonstrate how this is achieved in situations involving vehicle motion, the case for a stationary vehicle follows from results.

7.3.1 Demonstration 1

Two sensors on a vehicle are in a position to observe two features for purposes of localization using the decentralized algorithm implemented in Chapter 6. The scenario is illustrated in Figure 7.7. The action set in this case is given by

$$\mathcal{A} = \left\{ \begin{array}{l} a_1 = (t_a \rightarrow \text{sensor 2}, t_b \rightarrow \text{sensor 1}), \\ a_2 = (t_a \rightarrow \text{sensor 1}, t_b \rightarrow \text{sensor 2}) \end{array} \right\}.$$

Figure 7.7 (a) shows assignment (i) corresponding to a_1 and (b) shows assignment (ii) corresponding to a_2. The vehicle is stationary up until

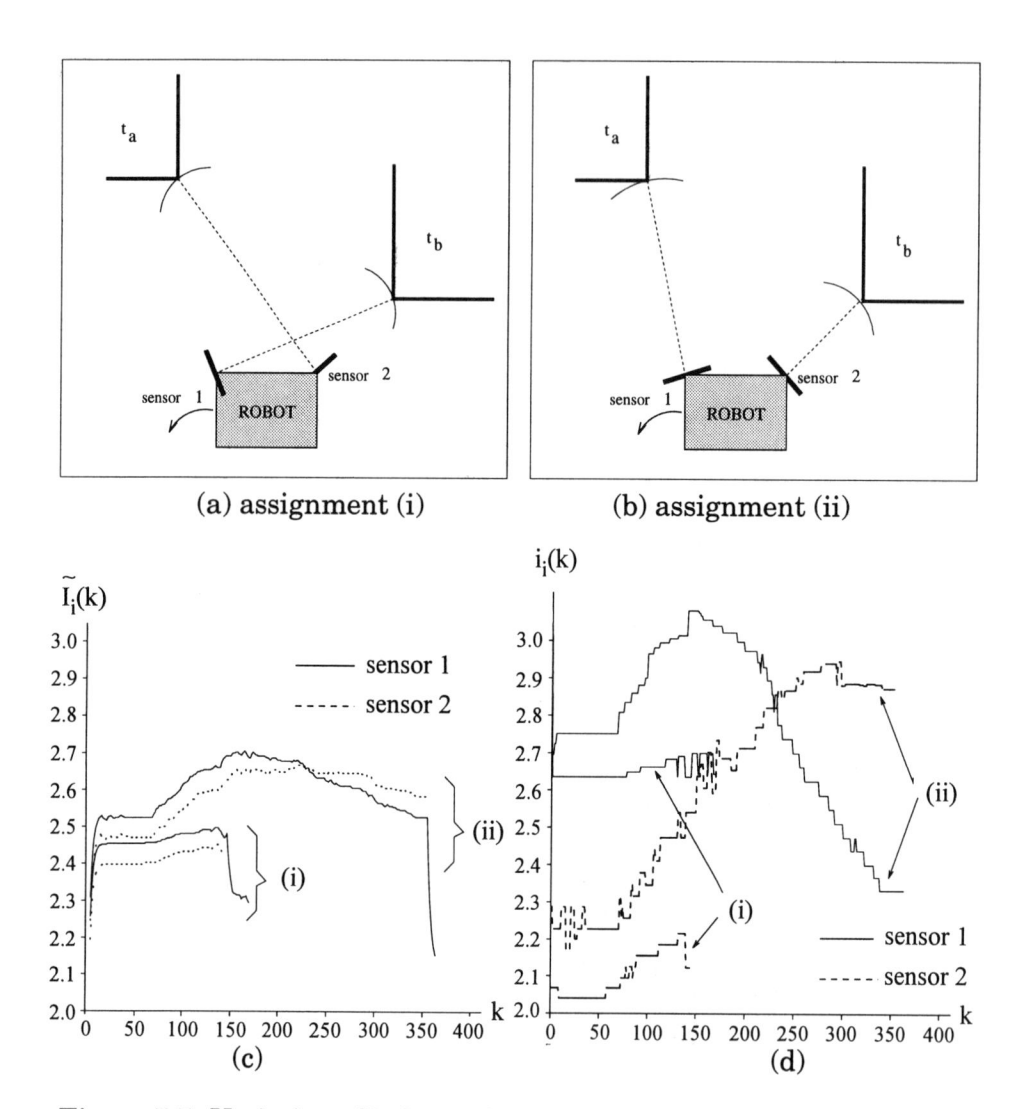

Figure 7.7: Variation of information quantities with two sensor-feature assignments while the vehicle is rotating counter-clockwise. (a) illustrates assignment (i) and (b) assignment (ii). Shown in (c) is the partial posterior information $\tilde{\mathbf{I}}_i(k)$ for each of the sensors corresponding to each assignment. Shown in (d) is the observation information $\mathbf{i}_i(k)$ for each sensor corresponding to each assignment. The run in assignment (ii) is shorter because at approximately $k = 150$ sensor 1 loses its feature as it moves out of view.

7.3 Sensor-Feature Assignments

time-step $k = 75$, after which it starts to rotate. Considering time prior to rotation, the following can be observed from (c) by considering partial information up to k;

$$(\tilde{\mathbf{I}}_1(k), a_2) > (\tilde{\mathbf{I}}_1(k), a_1) \quad \text{and} \quad (\tilde{\mathbf{I}}_2(k), a_2) > (\tilde{\mathbf{I}}_2(k), a_1).$$

By considering observation information at time k, the following can be observed from Figure 7.7 (d);

$$(\mathbf{i}_1(k), a_2) > (\mathbf{i}_1(k), a_1) \quad \text{and} \quad (\mathbf{i}_2(k), a_2) > (\mathbf{i}_2(k), a_1).$$

From Equation 4.57 the optimal action is clearly obtained as

$$\hat{a} = \arg\max_a \sum_j \tilde{\mathbf{I}}_j(k),$$

or by considering observation information

$$\hat{a} = \arg\max_a \sum_j \mathbf{i}_j(k).$$

In both cases $\hat{a} = a_2$. It can also be noticed that with both a_1 and a_2, sensor 1 consistently obtains more information than sensor 2 while the vehicle is stationary. At time-step $k = 75$ the vehicle starts to rotate and with both assignments, the information starts to increase. However, with assignment (i), sensor 1 soon loses its feature as it moves out of view and the tracking process is terminated. With assignment (ii) tracking continues. At time $k = 150$, the trend changes and sensor 1 experiences a drop in the information content of its observations as shown in (d). On the other hand sensor 2 continues to obtain increased information in its observations. At time $k = 250$, sensor 2 starts to obtain more information from its observations than sensor 1. This is reflected in the partial information content $\tilde{\mathbf{I}}_i(k)$ in (c) where that of sensor 1 falls below that of sensor 2. Because observation information is assimilated in the data fusion process, this results in *both* the partial information content metrics falling but with that of sensor 1 falling faster. It can also be seen that sensor 2 ends with the higher partial posterior information compared to the time prior to $k = 225$ when sensor 1 has the higher partial posterior information. Therefore, in terms of information partial posterior information, for this example, the optimal sensor-feature assignment starts as and remains assignment (ii) (that is, action a_2).

7.3.2 Demonstration 2

In this demonstration, the vehicle moves in a straight line as illustrated in Figure 7.8 (a) while each of the sensors tracks one of two features. The action set considered in this case, is given by

$$\mathcal{A} = \left\{ \begin{array}{l} a_1 = (corner \rightarrow \text{sensor 2}, cylinder \rightarrow \text{sensor 1}) \\ a_2 = (corner \rightarrow \text{sensor 1}, cylinder \rightarrow \text{sensor 2}) \end{array} \right\},$$

where in Figure 7.8, a_1 and a_2 are assignments (i) and (ii) respectively. The two possible assignments give rise to the results in (b) and (c). It can be seen that assignment (ii) results in higher posterior information. Although in assignment (ii) the sensors collectively obtain more information in their observations, sensor 2 in assignment (i) obtains increasingly higher information from its observations $\mathbf{i}_2(k)$ which start to raise the information $\tilde{\mathbf{I}}_1(k)$ and $\tilde{\mathbf{I}}_2(k)$ of assignment (i).

7.3.3 Demonstration 3

An optimal sensor-feature assignment in localization should allow the sensor system to obtain the most information possible about the location of the vehicle. To demonstrate this, we illustrate how the global posterior information $\mathbf{I}_i(k)$ is maximized through management decisions. In this demonstration, there are two features which can be used for localization by a vehicle with two Tracking Sonars. The action set under consideration is given by

$$\mathcal{A} = \left\{ \begin{array}{l} a_1 = (\text{corner} \rightarrow \text{sensor 2}, \text{corner} \rightarrow \text{sensor 1}), \\ a_2 = (\text{edge} \rightarrow \text{sensor 2}, \text{edge} \rightarrow \text{sensor 1}), \\ a_3 = (\text{corner} \rightarrow \text{sensor 2}, \text{edge} \rightarrow \text{sensor 1}), \\ a_4 = (\text{edge} \rightarrow \text{sensor 2}, \text{corner} \rightarrow \text{sensor 1}), \\ a_5 = (\text{edge} \rightarrow \text{sensor 1}) \end{array} \right\}.$$

The environment is illustrated in Figure 7.9 and shows the different sensing strategies given in the action set. In Figure 7.10 the global information from the information filter for each of the different strategies is plotted. It can be seen that the strategy corresponding to action (iii) results in the most information being obtained about vehicle location. Therefore, this action is chosen as the most optimal one. The maximization to obtain this optimal action is based on the algorithm communicating expected utilities in the form of partial information $\tilde{\mathbf{I}}_i(k)$ as discussed in Chapter 4.

7.3 Sensor-Feature Assignments 215

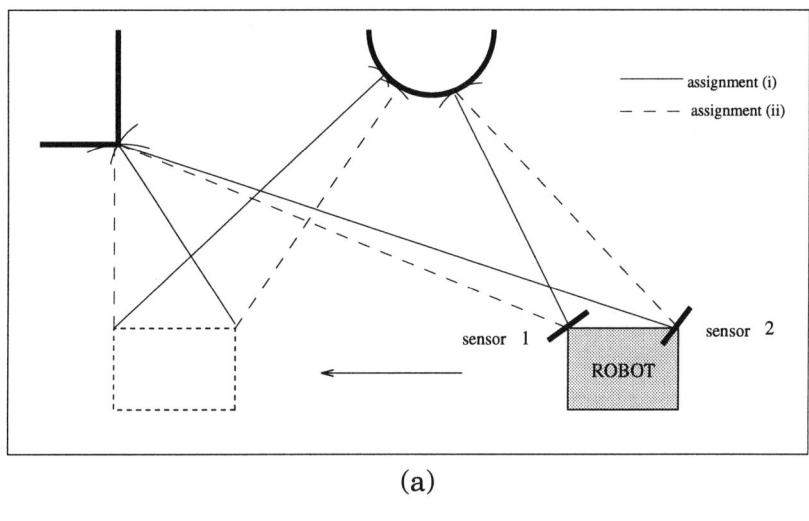

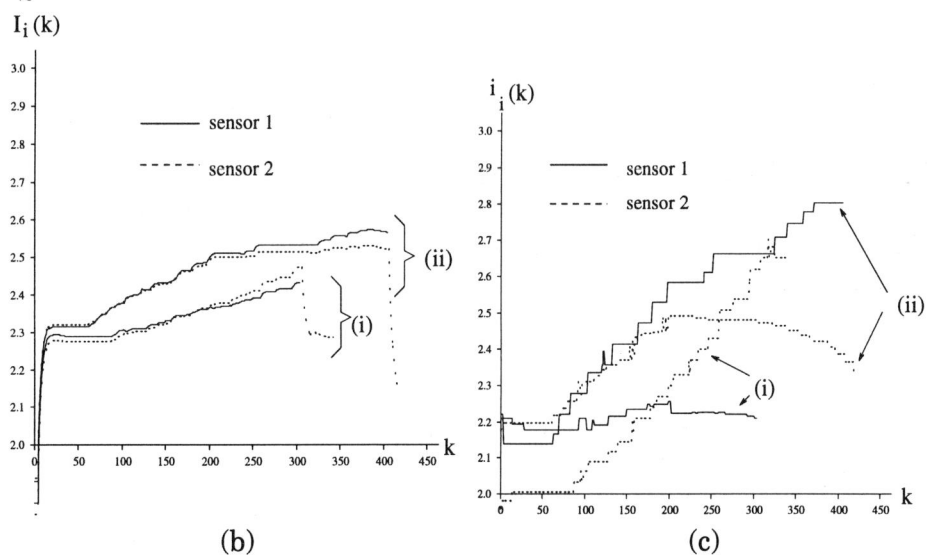

Figure 7.8: Variation of information quantities with two alternate sensor-feature assignments while the vehicle moves as indicated in (a). Illustrated in (a) are the assignments (i) and (ii). Shown in (b) is the partial posterior information $\tilde{\mathbf{I}}_i(k)$ for each of the sensors corresponding to each assignment. Shown in (c) is the observation information $\mathbf{i}_i(k)$ for each sensor corresponding to each assignment.

(a) action a_1

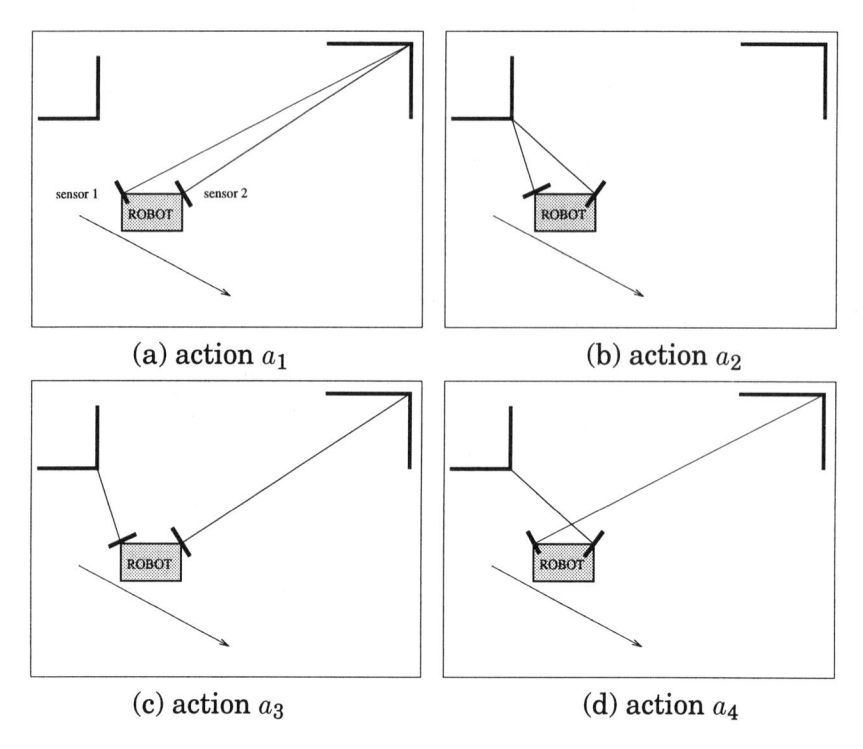

(b) action a_2

(c) action a_3

(d) action a_4

Figure 7.9: Various sensing strategies for a vehicle with two sensors and able to track two features. (a) assignment (i) (b) assignment (ii) (c) assignment (iii) and (d) assignment (iv). Plot (v) corresponds to only using sensor 1 to track the edge.

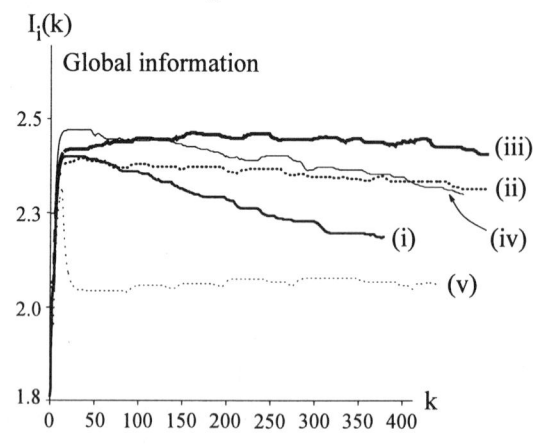

Figure 7.10: Total global information corresponding to each of the actions in Figure 7.9. Plot (v) corresponds to only using sensor 1 to track the edge.

7.4 Sensor Hand-off and Cueing

In the context of vehicle navigation using Tracking Sonars, sensor hand-off and cueing refers to the notion of a sensor currently observing a particular feature handing over observation of it to another sensor that has a higher utility for tracking that feature. This amounts to making decisions about which sensor is best suited to a given tracking task at any given point in time and, consequently, allowing that sensor to perform the tracking task. We demonstrate this for both the localization and the feature classification algorithms.

7.4.1 Demonstration 4

Consider two sensors on a vehicle, tracking the same feature for purposes of localization. In the first instance, the vehicle moves in straight line past the feature. In the results of Figure 7.11, we show the information quantities using two location covariance models with the variances as detailed in the figure caption. From the results in Figure 7.11 (a), it can be seen that with model 2, the hand-off point is difficult to distinguish if only partial information $\tilde{\mathbf{I}}_i(k)$ is considered. However, if observation information $\mathbf{i}_i(k)$ is also considered, the hand-off point is easily distinguishable with both models. The significance of this is that, if metrics which determine whether actions are worth implementing, such as described Section 4.5.2, are used based on a fixed threshold level, the information to be gained from management may fall below the threshold level. This means that the management decisions may not be implemented (see Equation 4.74). For this reason, *both* observation information $\mathbf{i}_i(k)$ and partial information $\tilde{\mathbf{I}}_i(k)$ must be considered for hand-off and cueing.

Figure 7.12 shows hand-off for a vehicle moving back and forth while two sensors track the same feature. Again the observation information $\mathbf{i}_i(k)$ in (b) shows the hand-off points quite distinctly, whereas the partial information $\tilde{\mathbf{I}}_i(k)$ in (a) does not. In (b) it can be seen that from the start to $k = 200$, sensor 1 has the higher utility and between $k = 200$ and $k = 400$ sensor 2 has the higher utility. After $k = 400$ sensor 1 again has the higher utility.

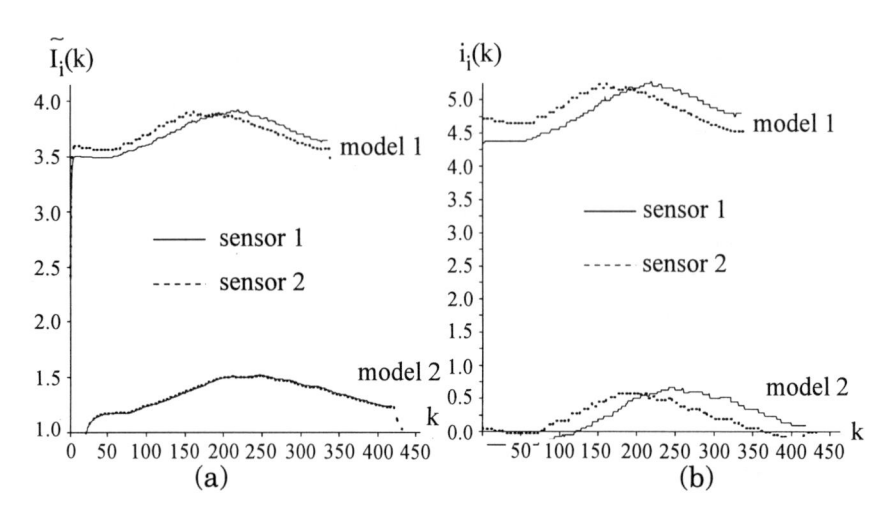

Figure 7.11: Partial and observation information used for hand-off while vehicle motion is in a straight line. Different parameters resulting in different location covariance models **R** are used, that is, model-1 is $\sigma_\theta^2 = \sigma_r^2 = 0.0001$ and model-2 is $\sigma_\theta^2 = 0.01, \sigma_r^2 = 0.005$. Hand-off occurs at $k = 210$.

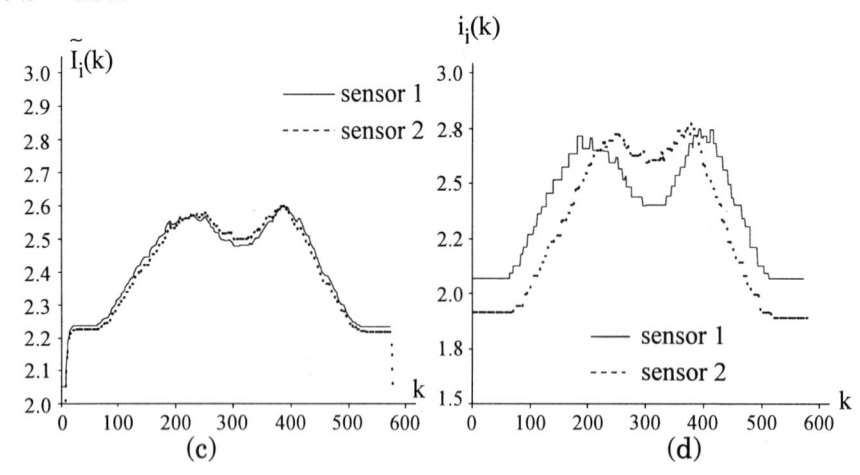

Figure 7.12: Partial and observation information used for hand-off as vehicle moves back and forth. Hand-off occurs at times $k = 200$ and $k = 400$.

7.4 Sensor Hand-off and Cueing

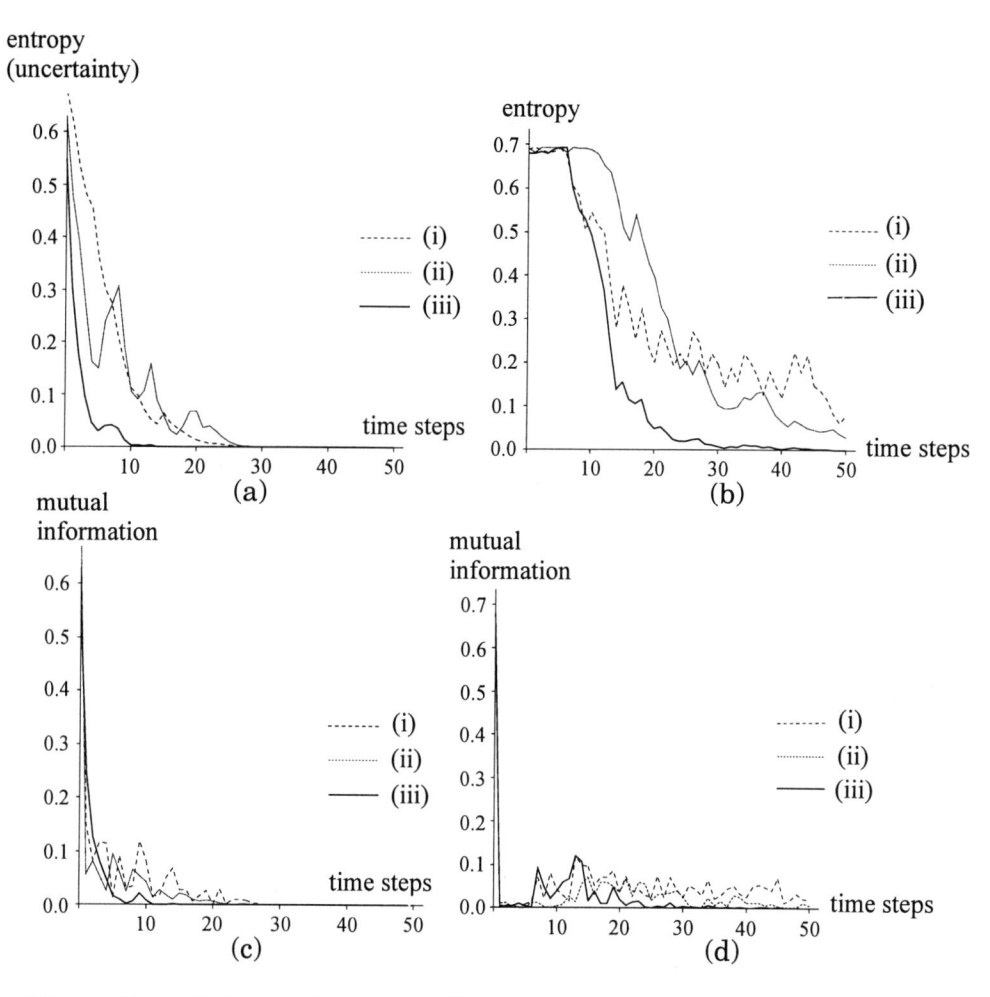

Figure 7.13: Information values for two sensors running the decentralized classification algorithm while tracking the same feature. (a) and (b) show the entropy ($-\tilde{\mathbf{I}}_i(k)$) (c) and (d) the mutual (observation) information $\mathbf{i}_i(k)$. In these (i) is sensor 1 information (ii) is sensor 2 and (iii) is the assimilated (fused) information. Two classification observation models are used, model 1 in (a) and (c) and model 2 in (b) and (d).

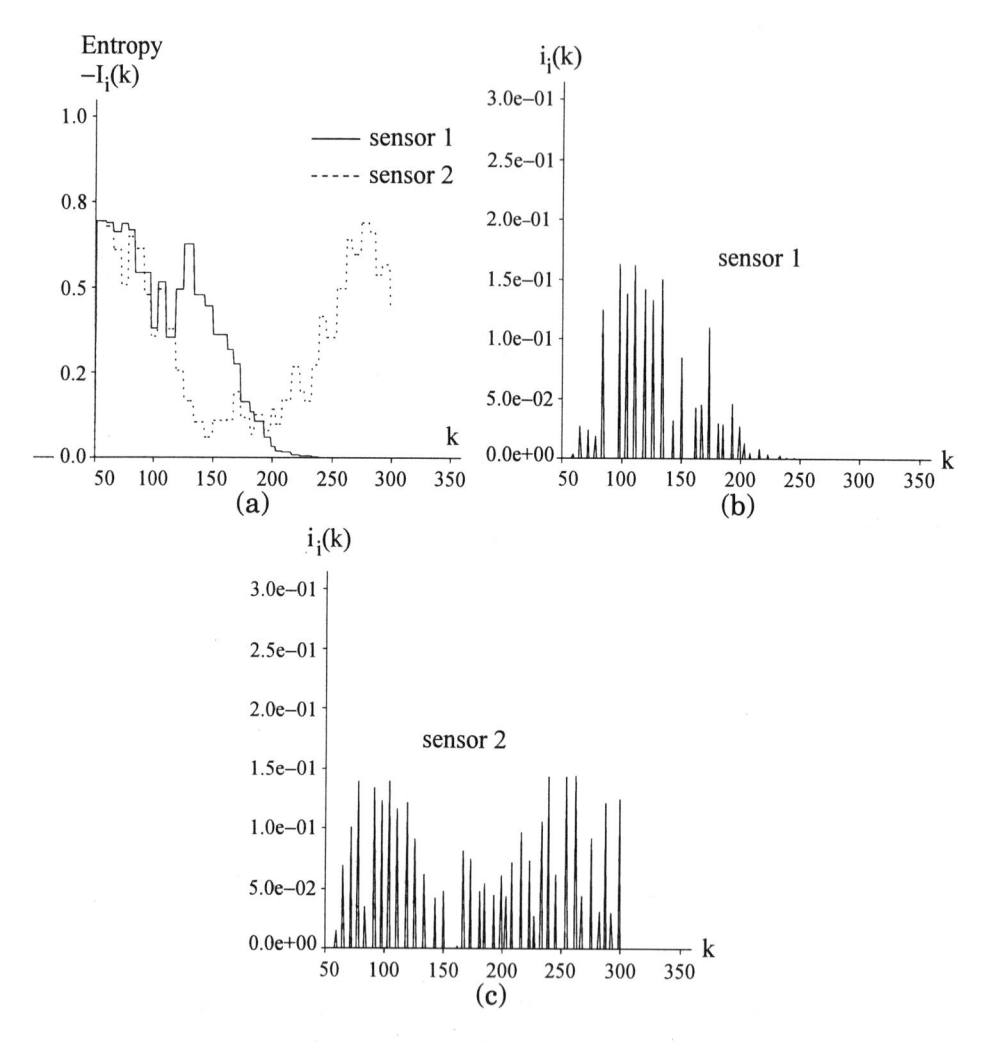

Figure 7.14: Partial information at each sensor for two sensors classifying the same feature. (a) shows the partial entropy at each sensor, (b) shows the observation (mutual) information at sensor 1 and (c) observation information at sensor 2.

7.4.2 Demonstration 5

Similar hand-off results can are obtained when two sensors are classifying the same feature as the results in Figures 7.13 and 7.14 show. The difference between the localization and classification results is that for the classification, the partial entropy ($-\tilde{\mathbf{I}}_i(k)$) is more useful than the observation information $\mathbf{i}_i(k)$. This is because the observation information is somewhat unstable and fluctuates so much such that it can result in unwarranted and frequent hand-off and cueing. This is evident in the results of Figure 7.13 (c) and (d), and Figure 7.14 (b) and (c). Cost metrics such as those based on the value of information and a threshold (see Chapter 4, Section 4.5.2), can be used to damp out the effect on decisions due to the rapid fluctuations in the observations by introducing a sort of "decision dead-band".

7.5 Summary

Due to the small number of sensors involved in our implementation on *Joey*, namely 4, the time taken to compute the rational action is negligible. Therefore, the bargaining algorithm of Chapter 4 can be run to completion at no significant computational cost. However, there are several practical problems associated with effecting management decisions, such as the occasional failure to validate a newly acquired target following a hand-off decision. These are largely a result of the hardware limitations of the Tracking Sonars and are eliminated or at least significantly reduced with improved hardware (see Appendix B.4) and better track validation. The management implementations, of which results have been presented here, can be extended to systems with a large number of sensors and action sets. And since performing the maximizations required can be computationally significant, the bargaining algorithm of Figure 4.9 can be employed.

While the results presented have been for a "stationary-feature-moving-vehicle" arrangement, similar results are obtained when the Tracking Sonars are on a stationary platform and the features themselves are moving. In such a case, if the Tracking Sonars are replaced by CCD cameras this becomes like a surveillance system where management decisions can be used to make sensor-target assignments in a similar way. The surveillance system described by Rao [178] is an example of a system which could be managed as discussed here. Another potential application is in the area of tracking vision heads such as Yorick which has been developed at Oxford [192].

7.6 Bibliographical Note

At the time of publication, there was hardly any published references to distributed systems making use of normative techniques for decentralized sensor management. However, for descriptions of implemented sensor management systems, the reader is referred to the comparison and results presented by Popoli [172] for a centralized multi-sensor system. Waltz and Llinas [203] also describe and show some results of implemented systems using non-normative techniques and Blackman [26] shows the results for a "normative-like" centralized radar system.

8 Towards General and Robust Managed Data Fusion

*What we call the beginning is often the end
And to make an end is to make a beginning.
The end is where we start from.*

- T.S. Eliot

This monograph has addressed data fusion and sensor management in multi-sensor systems in general and decentralized systems in particular. The methods that we have presented have been demonstrated by the application to mobile robotics. While the application chosen attests to our own research inclinations, the variety of other potential applications is wide. It is befitting, in conclusion, that in addition to discussing some of the limitations of our work and areas for further research, we also discuss other methods and frameworks which can alternately be adopted.

8.1 Review of Significance and Limitations

Importance of a Consistent Approach. The development of a unified approach to data fusion and sensor management has in the past not been crucial to the development and implementation of multi-sensor systems. This is because in centralized and hierarchical systems, ad hoc methods can be made to perform well by taking advantage of the fact that decisions are referred to the central processor which has the benefit of global knowledge, thereby guaranteeing consistency. With the advent of distributed and decentralized systems, the consistency and coherence of data fusion methods becomes necessary due to the presence of several decision-makers. In this respect, based on justifiable axiomatic considerations, we have presented a single framework in which consistency and coherence are implicit. However, the effectiveness of this framework is dependent on an ability to model information from sensors and the outcomes of decisions probabilistically.

The methods presented can be further developed and refined through application to systems with large action sets and a large number of sensors and the performance evaluated. The application presented in this monograph has been such that the number of sensor involved is relatively small and so management performance in this respect cannot be fully evaluated.

Sensor Models and (Relevant) Information. The sonar model that has been presented is intended to highlight the importance of understanding and correctly modelling information from a sensor. Such models are a requirement of normative methods for sensor management and also necessary for the uncertainty models used in the data fusion algorithms. The Tracking Sonar model is extremely useful to sensing for mobile robotics and represents an novel approach to sonar data processing. This sensor model extends the concept of "focus of attention" largely exploited in vision systems, to standard sonar at a cost several orders of magnitude less than that of equivalent vision systems[1]. Therefore, the model presented for the sonar sensor illustrates how a sensor can be designed and used in a directed fashion so as to obtain information that can be used directly. This can be generalized to say that, the sensor demonstrates what can be achieved using a focus of attention capability, in terms of reducing the data association considerations and obtaining only that information relevant to the problem at hand.

Needless to say, the practical implementation of the Tracking Sonar is very much at the prototype stage leaving room for further improvements and applications.

Limits of Information-based Normative Sensor Management. The sensor management methods presented in this book are applicable to a variety of situations where autonomous sensors in a decentralized system are required to make non-conflicting choices concerning sensing

---

[1]Such a comparison with vision systems is, however, quite naive given the enormous information potential of vision systems which is several magnitudes greater than that of standard sonar based on Polaroid devices. However, the comparison is, at the present time, one worth making given the complexities surrounding the practical direct use of the information obtained from equivalent vision systems with a focus of attention capability.

actions in a manner that is "optimal" and rational. However, the formulation, as we have presented it, is dependent on an ability to compare and evaluate the utilities of all sensors in the system. This amounts to evaluating alternate sensing actions for their informativeness with respect to the state being inferred, i.e. "information-maximizing". While this is defensible for a system whose primary goal is perception, difficulties arise in the evaluation of actions not directly concerned with the perception task at hand, such as when reacting to contingencies e.g. avoiding an obstacle. It can be argued, nevertheless, that any action by a sensor can always be abstracted to its informational value. What becomes a challenge, however, is assigning value in the form of expected utilities to such actions so as to facilitate decision-making. Moreover, assigning such value requires a probabilistic interpretation of the outcomes of the actions. This places limitations on the extendibility and maintainability of the sensor management system because any additions or changes in the functionality of the system require a "proper" probabilistic interpretation.

8.2 On General Methods and Paradigms

Due to the fact that applied research in decentralized autonomous sensor systems is fairly recent, some of the concepts we have presented are relatively new, and throughout this book we have endeavoured to highlight areas that can be explored further. In particular, since normative sensor management has not received much attention in the past, this warrants a discussion on directions for further research. Another consideration is that, while a Bayesian information-theoretic approach can be used to address problems in data fusion in general, it is by no means the only such paradigm. Therefore, a discussion of other, potential, generally applicable paradigms is beneficial.

8.2.1 Developing General Normative Management

Actions. The actions which can be evaluated using an information-theoretic approach, as presented in this book, are those directly related to sensing strategies whose value with respect to the inference process can be quantified as an expected utility. The ability to express value formally, in terms of expected utility, requires a probabilistic modelling of the outcome of a sensing operation. When actions and their outcomes cannot be subjected to such modelling, our approach, as presented here, becomes inadequate. It is important, therefore, to develop a normative method capable of handling a wide variety of sensing actions. Alternatively, a more tractable, and perhaps less all-encompassing, approach is to develop methods for handling classes of sensing actions. In this way, for example, actions related to reducing uncertainty about the state of nature would form a class.

Communication and computational costs. In the algorithms of Chapter 4, there is a need to communicate observations in order for each sensor to evaluate its own utility for tasks currently being carried out by another sensor. This can be avoided through a less rigorous method such as, for example, one in which utilities are based on segmenting the environment into grids in such a way that each grid has an associated utility[2]. While the discussion in Chapter 4 considered the computational and temporal costs of blindly pursuing a purely decision-theoretic solution, the cost of implementing actions was not considered. Clearly such costs in a real system, ought to have bearing on the choice of optimal action. A starting point in this direction can be based on the use of *cost functions* such those presented by Pearl [171]. Cost functions can also be used to model outcomes for a variety of actions. While the use of such techniques may be disconcerting to Bayesian purists who might call to question the normative status of the solution obtained, it can be argued that such cost functions can be developed axiomatically as has been done with utility functions. This is indeed the approach taken in the very recent work by Blackman for the management of agile radar [26]. Such cost functions provide a formal method for dealing with what may otherwise be an intractable problem in a strictly normative sense.

Large Distributed Networks. Our entire presentation of decentralized sensor management has been for fully connected systems. However, for large sensor networks, full inter-connection becomes impractical and hence the development of non-fully connected decentralized systems aimed at reducing the connectivity of the system while yielding results which approach those of fully connected systems. The methods that we have presented need to be extended to such systems. This can be done either by assuming the communication solutions presented

---

[2]This can be similar to the occupancy grids of Elfes [74] used in a different context.

8.2 On General Methods and Paradigms

by Grime and Durrant-Whyte [90] and applying them to the algorithms presented in this monograph or by adopting an alternate approach based on results from Team Theory [109]. Ho [108] gives a general discussion on the organisation of multi-sensor systems based on team theory. However, an approach which is closer to the work presented in this book, is the use of *sub-sampling* techniques amongst large groups of Bayesian decision-makers as discussed in the theoretical work by Weerahandi and Zidek [206] and Zidek [210]. In essence, sub-sampling means that from a large group of decision-makers, a smaller group with the most to gain takes charge of the decision process, and in this way, the decision problem amongst this smaller group becomes similar to that which we have presented.

8.2.2 Generally Applicable Paradigms

In proposing generally applicable paradigms for addressing managed data fusion in a consistent manner, the issues outlined in Chapter 1 can be used as criteria for suitability which must be met for a variety of multi-sensor architectures i.e.;

- the interpretation and representation of measurements and information from diverse sensors in a consistent manner,

- a capacity to address the fusion of information from several sources, and the inference and estimation of the relevant states,

- an ability to address sensor management and coordination rationally.

We now discuss these with respect to some of the approaches which can be adopted.

Bayesian Methods. It has been demonstrated that a Bayesian approach can form the basis of a consistent and generally applicable paradigm. However, difficulties may be encountered. Indeed applications do arise in which the age old Bayesian criticisms become problematic. Firstly, there are problems arising from the combination of priors and likelihoods, especially when priors determine the posterior and likelihoods are not informative (see discussion in Section 2.4.3). Secondly, the philosophical inconsistencies between ignorance and imprecision due to lack of information sometimes translate into practical difficulties when it comes to inference and estimation. And thirdly, in Chapter 2, it was assumed that each Bayesian was coherent (Assumption 1), and such coherence can, in practice, be guaranteed through data validation and consistency checks. However, in some applications data validation methods are not reliable and so such coherence and consistency cannot be guaranteed.

Belief Theory. An alternative paradigm which is amenable to generalization is the Dempster Shafer approach based on *belief theory*. In [203] Waltz and Llinas compare Bayesian and Dempster Shafer methods for data fusion. While belief theory does overcome some of the fundamental difficulties with Bayesian methods [203], it presents difficulties of its own such as numerical sensitivity and overestimation problems [3]. Belief theory can be developed and extended to address problems in sensor management. A set of hypotheses can be generated which is associated with a sensing action. An ambiguity measure associated with each hypotheses set can then be defined. It turns out that such a measure is a form of the entropy measure defined in Chapter 2. Sensor management decisions can then be made by evaluating each proposed action to find the one which minimizes the maximum ambiguity. This requires a predictive capability[3] and Hutchinson and Kak [114], for a particular problem, discuss how such prediction can be achieved using position transformations implicit in an object hypothesis. The problem therefore reduces to a search problem in which the search space is comprised of all the possible sensing strategies. Hutchinson and Kak [114] also present an algorithm for evaluating the next best sensing action. The search problem can be computationally significant and, in practice, heuristics have been used to reduce the necessary computations. The development and use of such heuristics may be difficult for some applications. Extension of the belief theory approach to decentralized architectures appears to be plausible.

Geometric Methods. Geometric methods are also a good basis for a generally applicable paradigm. In Chapter 2 we gave a geometric interpretation of uncertainty which can be used directly as a basis for fusion methodologies. In [161], Nakamura shows how multi-sensor fusion can be addressed using geometric methods with results similar to

---

[3]Prediction in the approach presented in this monograph is done using the predictive nature of the information filter.

those obtained using Bayesian methods. Geometric methods can be extended to sensor management and planning. Sensing actions can be planned and coordinated based on actively reducing geometric uncertainty volumes. The uncertainty of a state $\mathbf{x}$ is represented as an ellipsoid; $\mathcal{E}(\mathbf{x}) = d\mathbf{x}^T W_\mathbf{x} d\mathbf{x} \leq 1$, where $d\mathbf{x} \in d\mathcal{X}$, and $W_\mathbf{x}$ represents a weighting which determines the shape and size of the ellipsoid. The optimality of the sensor planning can then be measured by a comparison of the actual ellipsoid $\mathcal{E}(\mathbf{x})$ with the desired one $\mathcal{E}_d(\mathbf{y})$, where $\mathbf{y}$ is the desired output. Since the actual ellipsoid is a function of the controllable sensing parameters, a parametric constraint-based approach can be developed to iteratively obtain the appropriate control parameters. A preliminary discussion of sensor planning in this way, is presented by Lee and Schenker [130] as an extension to the SKC paradigm developed in [131]. Implicit in Lee and Schenker's approach is an ability to quantitatively model and state the sensing goals in terms of a desired error ellipsoid. Mathematically, geometric methods lend themselves easily to a variety of architectures, however, in practice this has yet to be explored fully.

8.3 Robust and Fully Functional Systems

While the development of theoretical frameworks and their validation through laboratory demonstrations is the stuff of academic research, considerably more work is required in order to implement fully functional and robust data fusion systems. It is thus important to discuss areas of importance in practical and fully functional systems.

8.3.1 Robustness

Of paramount importance for any practical system, is the issue of robustness which has not been addressed in great detail in this monograph. For systems implementing data fusion, robustness can be discussed on several levels. (Section 2.4.3 has already discussed Bayesian robustness.).

The architectures implementing decentralized data fusion are, in essence, distributed parallel systems. Implicit in decentralized systems is the communication and assimilation of information in the form of estimates etc. This is usually implemented as parallel communication on the parallel architecture. The subject of parallel communications comes with its own issues and problems of robustness such as deadlock, live-lock and synchronization. There are techniques available for verifying parallel communications such as the mathematical and formal *Communicating Sequential Processes* (CSP) [110]. Indeed this is the approach we have taken in [148], where the communication is verified for a decentralized architecture based on the Transputer. In this work a communication architecture, the Sensor Communicating Process Architecture (SPCA) is developed, verified and implemented on a decentralized system. Similar work verifying communication in sensor systems is shown in the work of Djian and Probert [64]. Such work illustrates that formal communication verification, although sometimes laborious, can be done. However, similar verification is yet to be done for non-fully connected systems such as those described by Grime and Durrant-Whyte [90].

Another robustness concern is data association and correlation. Detailed discussion of data association is outside the scope of this monograph, but nevertheless it is of paramount importance to the robustness of a practical data fusion system. As discussed in Chapters 3 and 6, Bayesian and filter techniques can be used to address data association. In some cases, however, it is necessary to address measurement validation at the sensor level and, often, this is done heuristically. Discussion of such methods is primarily sensor specific and is not discussed here[4]. An important advantage with using the focus of attention concept is that it alleviates the data association problem considerably.

The robustness of the methods presented in this monograph is also dependent on the nature and severity of the estimation errors. Estimation errors are dependent on a variety of issues such as initialization, filter convergence, errors due to non-linearities and also modelling errors and nature of assumptions made. These problems can be severe in non-linear systems. The effect of communication delays, synchronization and biases in estimates gain in importance as systems become non-fully connected. Other factors which affect robustness include the frequency with which management decisions are effected. Frequently implemented management decisions, such as in a system which requires the repositioning of sensors, introduce the likelihood of data association and correlation errors. This frequency can be determined by a threshold based on the value of information as discussed in Chapter 4. Sensor pa-

---

[4]Discussion of some of the techniques used can be found in the survey in Chapter 1 and the references found therein.

rameters such as those which govern the operation of a focus of attention sensor also affect overall robustness.

8.3.2 Mobile Robot Navigation

Making use of the Tracking Sonar and the algorithms for data fusion and sensor management, we have presented new concepts and techniques for realizing autonomous sonar navigation. However, the implementations that we have presented are but demonstrations and the OxNav vehicle is a research vehicle. Consequently, more work is required to produce robust and fully functional navigation systems. Our aim has been to introduce new approaches and to demonstrate what can be achieved. A first step towards fully functional systems is to improve the prototype design of the Tracking Sonar in order to make it more robust; firstly, by improving the hardware and, secondly, by developing a more stringent validation scheme for maintaining focus of attention. A more detailed quantitative analysis of performance is also required. The decentralized localization scheme needs to be more rigourously evaluated and tested. This entails considering various criteria such as; performance over extended periods, absolute accuracy of location information, and quantitative comparisons with other methods.

Ultimately, we envisage a fully functional vehicle making use of managed sensors navigating and building maps in a robust manner. A fully implemented system can be such that at any one time, each Tracking Sonar may be tracking a feature which has been matched against an environment map for localization or tracking an unknown feature and classifying it at the same time, for map-building purposes. Such a division of tasks among sensors would allow for simultaneous localization and map-building. Such work is underway on the OxNav vehicle project where the Tracking Sonars are being integrated into a sensor suite incorporating; inertial navigation sensors, a gyroscope, an infra-red sensor and encoders. This work integrates what has been presented in this monograph with that in Durrant-Whyte *et al* [31][18].

A   On Entropy and Information

A.1   Entropy of a vector distribution given covariance

For an n-dimensional vector $\mathbf{x}$ with a normal distribution, the PDF is given by

$$p(\mathbf{x}) = \mathbf{N}(\bar{\mathbf{x}}, \mathbf{P}) = |2\pi\mathbf{P}|^{-1/2} \, exp\left\{\frac{1}{2}(\mathbf{x}-\bar{\mathbf{x}})^T \mathbf{P}^{-1}(\mathbf{x}-\bar{\mathbf{x}})\right\}, \quad (A.1)$$

where $\bar{\mathbf{x}}$ is the mean of the distribution and $\mathbf{P}$ the covariance matrix. The entropy for this normal distribution is obtained as follows;

$$\begin{aligned}
E\{ln\,p(\mathbf{x})\} &= -\frac{1}{2}E\left\{(\mathbf{x}-\bar{\mathbf{x}})^T\mathbf{P}^{-1}(\mathbf{x}-\bar{\mathbf{x}}) + ln[(2\pi)^n|\mathbf{P}|]\right\} \\
&= -\frac{1}{2}E\left\{\sum_{ij}(\mathbf{x}_i-\bar{\mathbf{x}}_i)(\mathbf{P}^{-1})_{ij}(\mathbf{x}_j-\bar{\mathbf{x}}_j)\right\} - \frac{1}{2}ln[(2\pi)^n|\mathbf{P}|] \\
&= -\frac{1}{2}\sum_{ij}E\left\{(\mathbf{x}_j-\bar{\mathbf{x}}_j)(\mathbf{x}_i-\bar{\mathbf{x}}_i)\right\}(\mathbf{P})_{ij} - \frac{1}{2}ln[(2\pi)^n|\mathbf{P}|] \\
&= -\frac{1}{2}\sum_j\sum_i \mathbf{P}_{ji}(\mathbf{P}^{-1})_{ij} - \frac{1}{2}ln[(2\pi)^n|\mathbf{P}|] \\
&= -\frac{1}{2}\sum_j(\mathbf{P}\mathbf{P}^{-1})_{jj} - \frac{1}{2}ln[(2\pi)^n|\mathbf{P}|] \\
&= \sum_j \mathbf{I}_{jj} - \frac{1}{2}ln[(2\pi)^n|\mathbf{P}|] \\
&= -\frac{n}{2} - \frac{1}{2}ln[(2\pi)^n|\mathbf{P}|] \\
&= -\frac{1}{2}ln[(2\pi e)^n|\mathbf{P}|]. \quad (A.2)
\end{aligned}$$

From the above expression for information, the entropy is given by

$$E\{-ln\ p(\mathbf{x})\} = \frac{1}{2}ln[(2\pi e)^n|\mathbf{P}|]. \tag{A.3}$$

This results shows that all that is required to specify the entropy of a normal vector distribution is its length n and its covariance $\mathbf{P}$.

A.2 Non-informative priors and Entropy

Non-informative priors

A non-informative prior is a PDF that contains no information about the state $\mathbf{x}$. Finding such a distribution of the state $\mathbf{x}$, satisfying this condition is difficult when nothing is known concerning the space $\mathcal{X}$. A much used heuristic method is that due to Jeffreys [117] which is based on Fisher information, whereupon the prior is set to

$$p(\mathbf{x}) = |\ \mathbf{J}\ |^{1/2}, \tag{A.4}$$

where $\mathbf{J}$ is the expected Fisher information given by the definition in Equation 2.62 or Equation 2.63. The main attraction of this approach is that it is not affected in any way by restrictions on the state space $\mathcal{X}$ [23].

An approach commonly used when some information is available concerning the state is, ignoring philosophical inconsistencies, to assume an equivalence between "non-informativeness" and "least informativeness", so that the least informative distribution for $\mathbf{x}$ becomes the non-informative prior. The *Maximum entropy principle* states that the distribution which, for a given state gives the maximum entropy, can be used as the non-informative prior, thereby equating imprecision with ignorance. For a discrete distribution $\mathbf{x} = [X_1, X_2, \ldots, X_n]$, the maximum entropy distribution is obtained by assigning $p(X_i) = 1/n$, $\forall i$. The entropy for such a distribution is given by

$$h(p(\mathbf{x})) = -\sum \frac{1}{n}ln\left(\frac{1}{n}\right) = ln(n).$$

Since it can be shown that $h(p(\mathbf{x})) \leq ln(n)$ for all proper distributions $p(\mathbf{x})$ then indeed assigning equi-probabilities gives the maximum entropy distribution.

For continuous states, entropy is not bounded in this way and hence, in general, it has no maximum. However, if information in the form of moments is known about the space $\mathcal{X}$ this information can be used as constraints based on which a distribution maximizing entropy can be derived as we now outline:

Maximum entropy priors for continuous distributions. Moment constraints about the distribution $p(\mathbf{x})$ are usually given as follows

$$\int_{\mathcal{X}} f_i(\mathbf{x})p(\mathbf{x})d\mathbf{x} = \alpha_i. \tag{A.5}$$

What is then required is to find a maximum entropy distribution satisfying these moment constraints. Usually, for most continuous distributions only the first two moments are known, i.e.

$$\int \mathbf{x}p(\mathbf{x})\,d\mathbf{x} = \bar{\mathbf{x}} \quad \text{and} \quad \int \mathbf{x}^2 p(\mathbf{x})\,d\mathbf{x} = \mathbf{P} \tag{A.6}$$

The following are two approaches for obtaining the distribution maximizing entropy given these moment constraints:

- **Calculus Approach.** The Lagrangian for these constraints can be written

$$L \triangleq -\int p(\mathbf{x})\,ln[p(\mathbf{x})]d\mathbf{x} + \lambda_0 \int p(\mathbf{x})d\mathbf{x} + \sum \lambda_i \int f_i(\mathbf{x})p(\mathbf{x})d\mathbf{x}, \tag{A.7}$$

where the λ_is are Lagrangian multipliers. This can be maximized by differentiating and equating to zero

$$\partial L/\partial \mathbf{x} = -ln[p(\mathbf{x})] - 1 + \lambda_0 + \sum \lambda_i f_i(\mathbf{x}) = 0. \tag{A.8}$$

From this the form of the distribution which maximizes the entropy is obtained as

$$p(\mathbf{x}) = e^{\lambda_0 - 1 + \sum \lambda_i f_i(\mathbf{x})}, \tag{A.9}$$

where λ_i is chosen to maximize the constraints. Considering only the mean and covariance, Equation A.9 is written

$$p(\mathbf{x}) = e^{\lambda_0 + \lambda_1 \mathbf{x} + \lambda_2 \mathbf{x}^2}. \tag{A.10}$$

The result of this [42] shows that given the constraints, the normal distribution with the same first and second moments as given in the constraints, maximizes entropy.

A.3 The relationship between Fisher information and entropy

- **Using information inequalities.** Let $p(\mathbf{x})$ be the distribution for which a maximum entropy prior denoted $f(\mathbf{x})$ is required. Given that the mean of $p(\mathbf{x})$ is $\bar{\mathbf{x}}$ and the covariance is $\mathbf{P}$, by the constraints (Equation A.6), the mean and covariance of $f(\mathbf{x})$ are the same. From the definition of relative entropy [52] between two distributions, the following information inequality results

$$\int p(\mathbf{x}) ln\left[\frac{p(\mathbf{x})}{f(\mathbf{x})}\right] d\mathbf{x} \geq 0. \tag{A.11}$$

Simplifying and rearranging gives

$$0 \leq \int \left(p(\mathbf{x}) \, ln p(\mathbf{x}) - p(\mathbf{x}) \, ln f(\mathbf{x}) \right) d\mathbf{x}, \tag{A.12}$$

and using the definition of entropy yields

$$h(p(\mathbf{x})) \leq -\int p(\mathbf{x}) \, ln f(\mathbf{x}) d\mathbf{x}, \tag{A.13}$$

but $\int p \, ln f = \int f \, ln f$, since they yield the same moments (from the constraints), so

$$\begin{aligned} h(p(\mathbf{x})) &\leq h(f(\mathbf{x})) \\ &\leq \frac{1}{2} ln \left[(2\pi e)^n |\mathbf{P}| \right], \end{aligned} \tag{A.14}$$

using the result from Appendix A.1. Hence the normal maximizes the entropy for the given constraints.

A.3 On the relationship between Fisher information and entropy

Entropy and Fisher information are related by the log-likelihood function. Fisher information results in a MMSE estimate of the state $\mathbf{x}$ in that it provides a lower bound on the error in the MMSE estimate

$$\text{MSE} \geq tr(\mathbf{J}^{-1}). \tag{A.15}$$

This can be seen from the fact that each diagonal element of the estimation error covariance P_{ii} is the mean-squared error of the estimate of x_i, that is; $P_{ii} = E\{(\hat{x}_i - x_i)^2\} \geq (\mathbf{J}^{-1})_{ii}$, by the Cramer-Rao bound. Scharf

[188] shows that minimizing the estimated entropy is equivalent to an MMSE estimate.

Another consideration further strengthening the direct relation between entropy and Fisher information comes from consideration of the Asymptotic Equipartition Property (AEP) [52][1]. The AEP gives rise to the definition of the *typical set* A_ε^n. Formally, the typical set A_ε^n is the smallest volume set with probability greater or equal to $(1-\varepsilon)$ to the first order in the exponent. Informally, the typical set is the volume containing most of the probability. It can be shown that

$$Vol(A_\varepsilon^n) \leq 2^{n(h+\varepsilon)},$$

where h is the entropy associated with the corresponding PDF. From this we can write that $Vol(A_\varepsilon^n) \approx 2^{nh}$. Whereas entropy can be seen to be related to the volume of the typical set, Fisher information can be shown [52] to be inversely related to the surface area of the typical set. And so maximization of one is equivalent to minimization of the other. Defined as in Chapter 2, entropy is always a scalar and Fisher information always is a matrix for a vector **x**.

---

[1]Cover [52] gives a detailed development of the AEP for discrete and continuous r.v.'s, we make use of it here for illustrative purposes only.

B Differential Sonar Details

B.1 Sonar physical model

The range r to a feature (target) is obtained from

$$r = ct/2, \tag{B.1}$$

where c is the speed of sound in the transmission medium[1] and t is the measured time-of-flight. The receiving system detects the first echo which exceeds a threshold setting of the receiving circuit and ignores subsequent echoes. The principle is illustrated in Figure B.1 (a) which shows the observed time waveform. The impulse response can be analysed by separating the transmitter (T) and the receiver (R) and making the *far-field* approximation so that at the receiving element the spherical wavefront can be approximated by a plane wave. It has been shown by Kuc and Seigel [127] using Huygens principle that if angular inclination to the direction of propagation is α, the radius of the receiving aperture a, and the range r, the time function of the impulse response for a receiving only (R) aperture is given by

$$h_R(t, r, a, \alpha) = \begin{cases} k_1 \left(1 - \frac{c^2(t-2r/c)^2}{a^2 \sin^2 \alpha}\right)^{1/2}, & \text{for } \frac{2r-a \sin \alpha}{c} \leq t \leq \frac{2r+a \sin \alpha}{c} \\ & \text{and } 0 \leq |\alpha| \leq \pi/2 \\ \delta(t - 2r/c), & \text{for } \alpha = 0 \end{cases} \tag{B.2}$$

where

$$k_1 = \frac{2c \cos \alpha}{\pi \alpha \sin \alpha},$$

and δ is the Dirac Delta function [166]. The impulse response of Equation B.2 is illustrated in Figure B.1 (b). The impulse response of the

---

[1] In air this is 343.5 m/s at room temperature. This is also given by $\sqrt{\gamma RT}$.

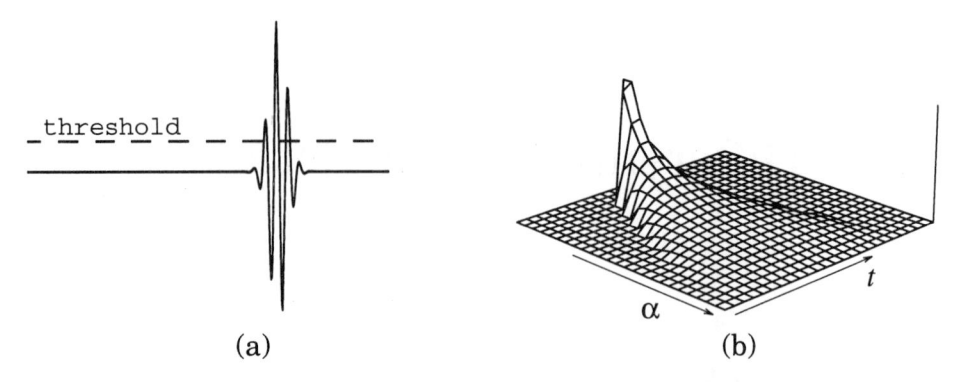

Figure B.1: (a) The observed time waveform (b) The impulse response of the receiver. The vertical axis shows the amplitude.

combination of a receiver and transmitter (T/R) is found to be a self convolution of the impulse response of the receiver h_R [127]

$$h_{T/R}(t,r,a,\alpha) = \int_{2r-a\ \sin\alpha/c}^{2r+a\ \sin\alpha/c} h_R(\tau,r,a,\alpha) * h_R(t-\tau,r,a,\alpha)\ d\tau. \quad (B.3)$$

For cylinders, whose radius is very small compared to the range, and edges the diffracted signal is more significant and so a factor $\sqrt{1/r}$ indicating the diminishing waveform due to diffraction with range r is included in the impulse response relationship. The beam has two distinct regions, the *Fresnel* (near) region and the *Fraunhofer* (far) region. The far region is defined by ranges $r > a^2/\lambda$. The beam emanating from the transducer can be described as shown by Morse and Ingard [158] in a rather complex analytical form. A simpler approximation is a Gaussian envelope [19]

$$p(\theta) = p_{max}\ exp\left[\frac{-2\theta^2}{\theta_0^2}\right], \quad \text{where}\quad \theta_0 = \sin^{-1}\left(\frac{0.61c}{fa}\right), \quad (B.4)$$

for the -30dB level.

It can be shown that the amplitude of the response decreases and its duration increases with the angle of inclination to the direction of propagation α. This means that for a given threshold setting in the receiving circuit, a feature will continue to be visible as the angle α is increased from zero to α_{max}. This is made possible by the fact that the beam-width on the Polaroid device is of the order of $10°$ at the -3dB point. Hence, the correct range to a feature will be measured for angles

B.2 Differential feature model

α, such that $-\alpha_{max} \leq \alpha \leq \alpha_{max}$, where α is measured from the direction of propagation to the target.

The effect of frequency on the angle α_{max} can be seen by considering the radiation characteristic of the beam pattern [134] emanating from the transducer given in Equation B.4. Figure B.2 shows the characteristic of Equation B.4 including the side-lobes. The figure illustrates the effect of decreasing the frequency of the transmitted signal. It can be seen that for a fixed threshold, the maximum α for reception is larger with a lower frequency f_2 than with a higher frequency f_1. The width

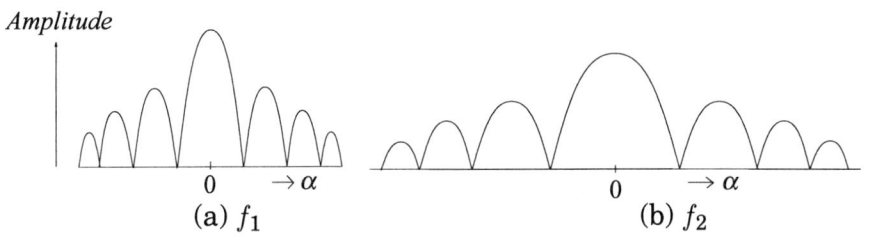

Figure B.2: The effect of varying frequency. In (a) and (b) $f_1 > f_2$.

of an RCD, therefore, depends on:

1. **Threshold of receiving circuit.** RCDs will appear wider with lower thresholds.

2. **Frequency.** Lowering the frequency will increase the RCD width for a given receiving threshold.

and to a lesser extent, the reflecting strength (acoustic impedance) of the target or obstacle.

B.2 Differential feature model

This section outlines, from a geometrical perspective, the form of the differential characteristic for each feature type, within the width of the RCD associated with that feature. In Figure B.3, a is the T/R device and b is the R device. For device a, a_t is the transmitter and a_r the receiver, for b we only have b_r. The differential is a consequence of the different path lengths.

Plane Model. By considering the geometry of Figure B.3 (a) we have

$$2p_1 = 2\left(r - \frac{d}{2}\sin(\alpha)\right). \tag{B.5}$$

240 **Differential Sonar Details**

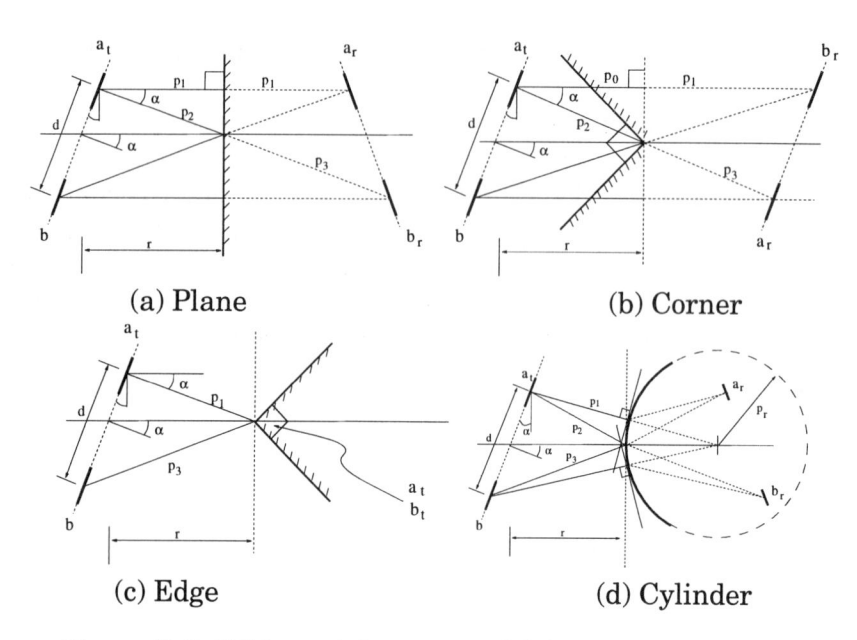

(a) Plane

(b) Corner

(c) Edge

(d) Cylinder

Figure B.3: Differential sonar model for various features.

$$p2 = \sqrt{\left(r - \frac{d}{2}\sin(\alpha)\right)^2 + \left(\frac{d}{2}\cos(\alpha)\right)^2} \quad \text{(B.6)}$$

$$p3 = \sqrt{\left(r + \frac{d}{2}\sin(\alpha)\right)^2 + \left(\frac{d}{2}\cos(\alpha)\right)^2}. \quad \text{(B.7)}$$

The differential path length δ_p is $\delta_p = (p_2 + p_3) - 2p_1$. Simplifying the above expressions this can be written

$$\delta_p = \left(\sqrt{r^2 + \frac{d^2}{4} - r\,d\,\sin(\alpha)} + \sqrt{r^2 + \frac{d^2}{4} + r\,d\,\sin(\alpha)}\right) - 2\left(r - \frac{d}{2}\sin(\alpha)\right). \quad \text{(B.8)}$$

Corner Model. For the corner in Figure B.3 (b), notice the reversal of a_r and b_r. For the reception at a_r the path length is given by $p_2 + p_3$ where

$$p_2 = p_3 = \sqrt{\left(r - \frac{d}{2}\sin(\alpha)\right)^2 + \left(\frac{d}{2}\cos(\alpha)\right)}. \quad \text{(B.9)}$$

B.2 Differential feature model

For reception at b_r the path length is given by

$$p_0 + p_1 = \left(r - \frac{d}{2}\sin(\alpha)\right) + \left(r + \frac{d}{2}\sin(\alpha)\right) = 2r. \qquad (B.10)$$

Thus, the differential is given by the path difference i.e.

$$\delta_c = (p_0 + p_1) - (p_2 + p_3) = 2\left[r - \sqrt{r^2 + \frac{d^2}{4} - rd\sin(\alpha)}\right]. \qquad (B.11)$$

Edge Model. For an edge the received signal is largely due to diffraction. We assume that the edge behaves as a rigid knife-edge [127] as shown in Figure B.3 (c). The virtual reflectors a_r and b_r coincide at the point of intersection as shown in the figure. For reception at a_r the path is given by $2p_1$ and that at b_r by $p_1 + p_3$. From Figure B.3 (c) we have that

$$p_1 = \sqrt{\left(r - \frac{d}{2}\sin(\alpha)\right)^2 + \left(\frac{d}{2}\cos(\alpha)\right)^2}, \qquad (B.12)$$

$$p_3 = \sqrt{\left(r + \frac{d}{2}\sin(\alpha)\right)^2 + \left(\frac{d}{2}\cos(\alpha)\right)^2}. \qquad (B.13)$$

The differential path length is $\delta_e = p_3 - p_1$. Simplifying using Equation B.12 and Equation B.13 gives

$$\delta_e = \sqrt{r^2 + \frac{d^2}{4} - rd\sin(\alpha)} - \sqrt{r^2 + \frac{d^2}{4} + rd\sin(\alpha)}. \qquad (B.14)$$

Cylinder Model. The path length for reception at a_r is given by $2p_1$. And that at b_r is given by $p_2 + p_3$. From the geometry of Figure B.3 (d)

$$p_1 = \sqrt{\left(\frac{d}{2}\cos(\alpha)\right)^2 + \left(r - \frac{d}{2}\sin(\alpha) + p_r\right)^2} - p_r, \qquad (B.15)$$

where p_r is the radius of the cylinder. Paths p_2 and p_3 are given by

$$p_2 = \sqrt{\left(r - \frac{d}{2}\sin(\alpha)\right)^2 + \left(\frac{d}{2}\cos(\alpha)\right)^2}, \qquad (B.16)$$

$$p_3 = \sqrt{\left(r - \frac{d}{2}\sin(\alpha)\right)^2 + \left(\frac{d}{2}\cos(\alpha)\right)^2}, \qquad (\text{B.17})$$

respectively. Thus the path difference is given by

$$\begin{aligned}\delta_{cyl} &= (p_2 + p_3) - 2p_1 \\ &= \left(\sqrt{r^2 + \frac{d^2}{4} - r\,d\,\sin(\alpha)} + \sqrt{r^2 + \frac{d^2}{4} + r\,d\,\sin(\alpha)}\right) \\ &\quad - 2\left(\sqrt{\left(\frac{d}{2}\cos(\alpha)\right)^2 + \left(r - \frac{d}{2}\sin(\alpha) + p_r\right)^2} - p_r\right).(\text{B.18})\end{aligned}$$

The differential for the cylinder approximates that for the plane as p_r goes to infinity. This can be deduced from a consideration of p_1, given by Equation B.15, that

$$p_1 \xrightarrow{p_r \to \infty} \left(r - \frac{d}{2}\sin(\alpha)\right), \qquad (\text{B.19})$$

which is the p_1 for a plane (from Equation B.5). Since p_2 and p_3 are not affected by changes in p_r, Equation B.18 and hence the differential, approximates that for the plane. When the radius is small p_1, as given by Equation B.15, becomes

$$p_1 \xrightarrow{p_r \to 0} \sqrt{\left(\frac{d}{2}\cos(\alpha)\right)^2 + \left(r - \frac{d}{2}\sin(\alpha)\right)^2}, \qquad (\text{B.20})$$

which is the same as p_1 for the edge (Equation B.12). Hence as the radius of the cylinder goes to zero the differential of Equation B.18 approximates that for an edge. Thus, for any cylinder the true differential lies between that for the edge and that for the plane.

From the geometrical forms of the differential given above, its can be shown that the differential characteristics for all four feature types are similar albeit with slight variations. These variations can be attributed to the asymmetry of the device arrangement and also to the geometrical shapes of the features. The variations are, in practice, hardly noticeable. However with more accurate measurement the variations may shed some light on feature classification based on the differential characteristic.

Other than physical compactness, there are other considerations in the choice of base-line d. A very small d results in a very small gradient for the differential which may be difficult to measure accurately. Thus there is a need to reconcile by way of compromise, the gradient of the differential, hardware timing capabilities and the resulting timing discrimination. Another important consideration is the ability to discriminate between features which may be close together. Kleeman and Kuc [124] show for a similar sensor that the minimum feature separation s for discrimination is given by

$$s_{min} = \frac{d}{2} \sin\left(\frac{\omega}{2}\right), \tag{B.21}$$

where ω is the beam-width. Our present prototype hardware provides a precision of $3.2\mu s$ per count for the measurement of the differential TOF. And we have chosen a base-line of 0.05m for the operational range of 0.45m to 5.00m.

B.3 Tracking Sonar performance and limitations

Speed of relative motion

The limited sonar firing rate coupled with the fact that the motor requires a finite time to effect correctional motion, places a limit on the data rate of the sensor. The maximum data rate of the sensor is approximately 30Hz. Relative motion with regard to this sensor, can be decomposed into *angular* motion and *linear* motion along the normal to the feature being tracked. This particular decomposition of the relative motion is a result of the fact that the tracking is effected by correcting for the angular displacement and updating the corresponding range to the feature as shown in the algorithm of Figure 5.6. Assuming the worst case scenario that every return requires correctional-motion, the angular speed of the relative motion is limited to a maximum of approximately 0.6 radians per second while tracking an RCD of width 0.17 radians using the simple algorithm of Figure 5.6. However, this maximum speed can be increased to about 0.9 radians per second, through the adjustment of filter parameters, dead-band settings and a variable size of the correctional move step, optimized for a particular environment.

Of secondary importance is the linear speed along the normal to the target. In addition to the sonar firing rate, the limiting factor here is the nature of the range validation and update scheme. Assuming no relative angular motion and with a simple range validation and update scheme, the maximum linear speed, along the normal to the feature, giving satisfactory results is approximately 2.5 metres per second.

Nature and range of the tracked feature

If the feature is a plane, corner, edge or cylinder, then the differential characteristic is well-defined in the sense that it approximates a linear transition such as that predicted from geometrical considerations (see Appendix B.2). Satisfactory tracking performance is obtained if the feature has a well-defined differential characteristic. If it is known *a priori* that the features to be tracked do not have well-defined characteristics, then the differential observation noise model σ_δ^2, can be increased to take account of the increased observation noise. Irregular surfaces such as human bodies have ill-defined differentials, however, these can be tracked with limited success at short ranges, 0.45m to 2.00m and low relative angular speeds of the order of 0.35 radians per second. At short range (0.45m to 1.5m), firing rates can be as high as 30Hz and at ranges of about 4m the maximum firing rate giving satisfactory results is about 25Hz. For a fixed base-line d, the ability to measure the differential accurately deteriorates with range, a result which can be obtained by analysing the variations and uncertainty in the differential with variations in range.

The differential sonar does not directly provide information which differentiates the different target types. The information obtained with each return (see Equation 5.2) in itself gives no indication as to the nature of the target. However, as an indirect consequence of the ability to fixate on a given feature, a method based on the classification algorithm can be developed which results in classification of the feature.

Hardware limitations

With the current hardware, the differential can only be measured to a precision of $3.2\mu s$ up to a maximum value of $15 \times 3.2\mu s$. The smallest step possible with the motor is 0.047 radians which is crude given that the minimum RCD width which can be tracked is 0.17 radians. However, these limitations can be overcome with improved hardware designs.

B.3 Tracking Sonar performance and limitations

Operation together with other Tracking Sonars

An important consideration in the context of data fusion, is the performance of the Tracking Sonar in an environment where other Tracking Sonars are operating. When such sensors are used to track different features which are not in close proximity, the performance attainable is the same as in the preceding discussion for a single sensor. However, when the sensors are used to track the same feature or features in very close proximity ($\approx$ 0.2m apart), interference may occur if very high data rates of the order of 30Hz are maintained. This is because sensors begin to receive echoes from each other causing erroneous differential measurements. However, reducing the data rates to a maximum of approximately 20Hz gives reliable performance. Such a reduction in maximum data rates reduces the angular speeds of relative motion possible. As before, various optimizations are possible, such as introducing an "interlacing" scheme whereby for example, sensors do not fire synchronously for. Such techniques can be used to raise the maximum rates albeit to levels below the maximum attainable with a single sensor; for instance, with simple hardware rates of about about 26Hz can be reached using two such sensors in close proximity (0.3m) tracking the same feature at a range of 0.8m.

Feature discrimination

With the tracking implemented as in Figure 5.6, the tracking of a feature is somewhat indiscriminate. This is due to the fact that the sensor essentially tracks the null of the differential and updates the range accordingly. Thus if another feature coincides spatially with a feature being tracked, it is possible for the sensor to switch to the new feature apparently non-deterministically. It is possible to overcome this with a more stringent validation scheme and for this, increased accuracy is required in the parameters which are measured and in the positioning of the motor. In this way the basic algorithm can be extended to allow for momentary obscuring of the feature. This is desirable in cluttered environments where targets could become partially or momentarily obscured. The basic algorithm of Figure 5.6 assumes that there is always a clear path between the feature and the sensor thus limiting such an implementation to environments where this can be guaranteed.

Dependence of variances on the measured quantities

Within the working range of 0.045m to 4.00m, for a given set of sensor parameters ζ, the variances σ_r^2 and σ_θ^2 depend on the measured quantities themselves. This dependence can be modelled by modifying the nominal variances using functions which depend on the measured quantities r and θ. These functions can be determined empirically by obtaining several results such as those of Table 5.1 at various values of r and θ and finding functions which approximately fit the variations observed. For this sensor, this can be approximated using the functions

$$g_r(r) = 1 - 2(\sqrt{r})e^{-r}, \quad g_\theta(\theta) = (1 - e^{\frac{-(\theta-2)^2}{5}}). \tag{B.22}$$

for the σ_r^2 and σ_θ^2 respectively. These are then used in Equation 5.18. In this way the covariance $\mathbf{R}[k, \mathbf{z}(k)]$ becomes a *location covariance*.

B.4 Hardware

B.4.1 Generic decentralized sensor node architecture

Decentralized architectures can be implemented using parallel processing hardware. The group at Oxford has developed a generic architecture for decentralized system based on the Transputer processor. With this architecture, a sensor node has all the requisite communication links and interface logic necessary. The architecture features:

- A size TRAM1 T800 or T805 Transputer module, with full floating-point ALU.

- 4 External Communication links, with data rates of 10/20 MBits per second.

- Input/Output handled by C011 serial to parallel interface chips.

The entire system of communication and interface logic can be integrated onto a single chip as shown in Figure B.4. This architecture has been used in application ranging from distributed process plant monitoring to mobile robotics, as reported in [18][91][39].

B.4 Hardware

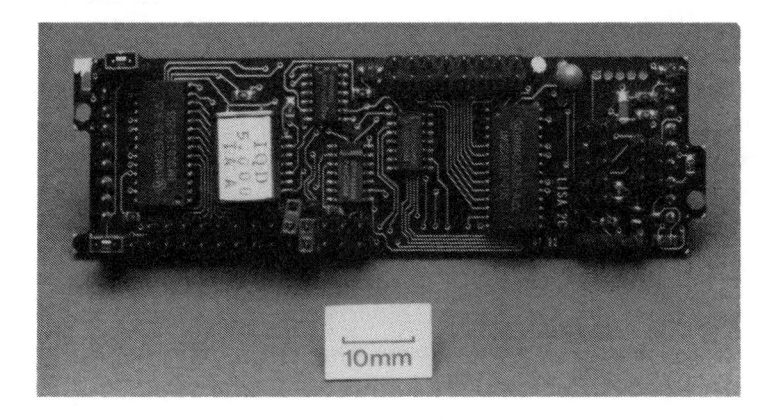

Figure B.4: Generic Transputer-based decentralized sensor node hardware.

B.4.2 Differential sonar

The differential sonar module integrates an entire sensing subsystem based on the generic sensor architecture described above. The hardware for the module was designed by I. Treherne at Oxford. The sensor module features:

- Communication using INMOS serial links at 10Mbaud or 20Mbaud.
- Option for on-board processor size 1 TRAM format eg Transputer.
- Motor driver motor circuitry.
- Sonar transceiver/receiver subsystems on board.
- Independent, fully isolated power supplies for sonar, motor and logic, sonar and motor power switchable.
- Optional communications watchdog.
- Application I/O (4 bits input + 4 bits output).

With each return, the differential sonar unit provides the following; $[r, \theta, |\Delta|, sign(\Delta)]$. These values are obtained as follows:

- **Range** r. The range of obtained by measuring the TOF as usual. The range can be resolved to an accuracy of approximately 1%.

- **Orientation** θ. The orientation is obtained from the direction in which the motor points the two Polaroids.

- **Differential** $|\Delta|$. The differential TOF is obtained by measuring the elapsed time between the primary reception and secondary reception. With current hardware, the differential TOF can be resolved to an accuracy of $3.2\mu s$.

- $Sign(\Delta)$. The sign of the differential TOF is given by a bit which notes the order of reception.

The above specification is clearly somewhat crude and this is has been to the detriment of the results obtained. An improved design was completed at the time of writing, which offered improvements of at least an order of magnitude in all the above.

Bibliography

[1] *New Scientist*, (1910):page 19, 29 January 1994.

[2] Abidi and Gonzalez. Introduction. In *Data Fusion In Robotics And Machine Intelligence*, pages 1–6. Academic Press, 1992.

[3] M.A. Abidi. Fusion of multi-dimensional data using regularization. In *Data Fusion In Robotics And Machine Intelligence*, pages 415–456. Academic Press, 1992.

[4] J. Alexander and J. Maddocks. On the kinematics of wheeled mobile robots. In I.J. Cox and G.T. Wilfong, editors, *Autonomous Robot Vehicles*. Springer-Verlag, 1990.

[5] P. Allen and R. Bajcsy. Two sensors are better than one: Example of vision and touch. In *Third Int. Symp. Robotics Research*, pages 48–55, Gouvieux, France, 1986. MIT Press.

[6] A. T. Alouani. Nonlinear data fusion. In *Proc. 29th CDC*, pages 569–572, Tampa, December 1989.

[7] F.P. Anderson. Visual algorithms for autonomous navigation. In *Proc. IEEE Int. Conf. Robotics and Automation*, page 856, 1985.

[8] R. L. Andersson. *A Robot Ping-Pong Player*. MIT Press, 1988.

[9] F.J. Anscombe and R.J. Aumann. A definition of subjective probability. *Ann. Math. Statist*, 34:199–205, 1963.

[10] K.J. Arrow. *Social Choice and Individual Values 2nd Ed*. John Wiley, New York, 1966.

[11] M. Bacharach. Group decisions in the face of differences of opinion. *Management Science*, 22(2):182–191, 1975.

[12] M. Bacharach. Normal bayesian dialogues. *J. American Statistical Soc.*, 74:837–846, 1979.

[13] J.G. Balchen and F. Dessen. Structural solution of highly redundant sensing in robotic systems. In *Highly Redundant Sensing in Robotic Systems*, pages 264–275. Springer-Verlag, Berlin, 1990.

[14] Y. Bar-Shalom. *Multi-Target Multi-Sensor Tracking*. Artech House, 1990.

[15] Y. Bar-Shalom. *Multi-Target Multi-Sensor Tracking Vol II*. Artech House, 1992.

[16] Y. Bar-Shalom and T.E. Fortmann. *Tracking and Data Association*. Academic Press, 1988.

[17] B. Barshan and H.F. Durrant-Whyte. Evaluation of a solid-state gyroscope for robotics applications. In *Submitted to IEEE Int. Conf. Robotics and Automation, Atlanta, U.S.A.*, 1993.

[18] B. Barshan and H.F. Durrant-Whyte. Inertial navigation system for a mobile robot. In *1st IFAC International Workshop on Intelligent Autonomous Vehicles, Southampton, U.K.*, 1993.

[19] B. Barshan and R. Kuc. Differentiating sonar reflections from corners and planes by employing an intelligent sensor. *IEEE Trans. Pattern Analysis and Machine Intelligence*, 12(6):560–569, 1990.

[20] O. Basir and H. Shen. A new approach for aggregating multisensory data. Technical report, Dept of Systems Design Eng. University of Waterloo, Ontario, 1992.

[21] G.E. Beck. *Navigation Systems*. Van Nostrand Reinhold, 1971.

[22] T. Berg and H.F. Durrant-Whyte. Model distribution in decentralized multi-sensor fusion. In *Proc. American Control Conference*, pages 2292–2294, 1991.

[23] J.O. Berger. *Statistical Decisions (second edition)*. Springer-Verlag, Berlin, 1985.

[24] S. Bier and P. Rothman. Intelligent sensor management for beyond visual range air to air combat. In *IEEE NAECON*, pages 264–270, 1988.

[25] S.S. Blackman. *Multiple Target Tracking with Applications to Radar*. Artech House, 1986.

[26] S.S. Blackman. Multi-target tracking with agile beam radar. In Y. Bar-Shalom, editor, *Multi-Target Multi-Sensor Tracking*, pages 237–268. Artech House, 1992.

[27] D. Blackwell and M. A. Girshick. *Theory of Games and Statistical Decisions*. Dover Publications, 1954.

[28] D.G. Bobrow. *Qualitative Reasoning About Physical Systems*. MIT Press, Boston MA, 1985.

[29] P. L. Bogler. Shafter-dempster reasoning with applications to multi-sensor target identification systems. *IEEE Trans. Systems Man and Cybernetics*, 17(6):968–977, 1987.

[30] J. Borenstein and Y. Koren. Obstacle avoidance with ultrasonic sensors. *IEEE J. Robotics and Automation*, 4:213–218, 1988.

[31] S. Borthwick, M. Stephens, and H. Durrant-Whyte. Position estimation using optical range data. In *Proc. Int. Symp. Intelligent Robots*, 1993.

[32] J.M. Brady, S. Cameron, H. Durrant-Whyte, M. Fleck, D. Forsyth, A. Noble, and I. Page. Progress towards a system that can acquire pallets and clean warehouses. In *Fourth Int. Symp. Robotics Research*, Santa Cruz, August 1987.

[33] J.M. Brady, H.F. Durrant-Whyte, H. Hu, J. Leonard, P. Probert, and B.S. Rao. Sensor based control of AGVs. *IEE Computing and Control Journal*, 1(1):64–71, 1990.

[34] R.A Brooks. Aspects of mobile robot visual map making. In *Second Int. Symp. Robotics Research*, Tokyo, Japan, 1984. MIT Press.

[35] C.M. Brown, H. Durrant-Whyte, J. Leonard, and B. Rao. Centralized and decentralized kalman filter techniques for tracking, navigation and control. In *Proc. DARPA Image Understanding Workshop*, pages 651–676. Morgan Kaufmann Inc., 1989.

[36] M.K. Brown. Feature extraction techniques for recognizing solid objects with an ultrasonic range sensor. *IEEE J. Robotics and Automation*, 1(4):191–205, 1985.

[37] R.G. Brown and P.Y.C Hwang. *Introduction to Random Signals and Aplied Kalman Filtering (2nd Ed)*. John Wiley, 1992.

[38] D.M. Buede and E.L. Waltz. Benefits of soft sensors and probabilistic fusion. In *SPIE Signal and Data Processing of Small Targets*, volume 1096, pages 309–320, 1989.

[39] T. Burke and H.F. Durrant-Whyte. Modular mobile robot design. In *1st IFAC International Workshop on Intelligent Autonomous Vehicles, Southampton, U.K.*, 1993.

[40] A. Cameron and H. Durrant-Whyte. Optimal sensor placement. *Int. J. Robotics Research*, 9(3), 1990.

[41] D.A. Castanon and D. Teneketzis. Distributed estimation algorithms for nonlinear systems. *IEEE Trans. Automatic Control*, 30(5):418–425, 1985.

[42] D. Catlin. *Estimation, Control, and the Discrete Kalman Filter*. Springer-Verlag, 1989.

[43] Z. Chair and P.K. Varshney. Optimal data fusion in multiple sensor detection systems. *IEEE Trans. Aerospace and Electronic Systems*, 22:98–101, 1986.

[44] Z. Chair and P.K. Varshney. Distributed bayesian hypothesis testing with distributed data fusion. *IEEE Trans. Systems Man and Cybernetics*, 18(5):695–699, 1988.

[45] K.C. Chang, C.Y. Chong, and Y. Bar-Shalom. Joint probabalistic data association in distributed sensor networks. *IEEE Trans. Automatic Control*, 31(10):889–897, 1986.

[46] R. Chatila. Mobile robot navigation: Space modelling and decisional processes. In *Third Int. Symp. Robotics Research*, Gouvieux, 1985.

[47] H. Chernoff and L. Moses. *Elementary Decision Theory*. Dover, 1959.

[48] C. Chong. Hierarchical estimation. In *2nd MIT/ONR CCC Workshop*, Monterey CA, 1979.

[49] C. Chong, S. Mori, and K. Chan. Distributed mutitarget multisensor tracking. In Y. Bar-Shalom, editor, *Multitarget Multisensor Tracking*. Artech House, 1990.

[50] A.G. Cohn. Approaches to qualitatitive reasoning. *Artificial Intelligence Review*, 3:177–232, 1989.

[51] Polaroid Corporation and Commercial Battery Division. *Ultrasonic ranging system*. 1984.

[52] T.M. Cover and J.A. Thomas. *Elements of Information Theory*. Wiley Series in Telecommunications, 1991.

[53] R.A. Cowan. Improved tracking and data fusion through sensor management and control. In *Data Fusion Symp.*, pages 661–665, 1987.

[54] I.J. Cox. Blanche: An autonomous robot vehicle for structured environments. In *Proc. IEEE Int. Conf. Robotics and Automation*, pages 978–982, 1988.

[55] I.J. Cox and J. Leonard. Probabilistic data association for dynamic world modeling:a multiple hypothesis approach. In *Intl. Conf. on Advanced Robotics. Pisa Italy*, 1991.

[56] I.J. Cox and G.T. Wilfong. *Autonomous Robot Vehicles*. Springer-Verlag, 1990.

[57] J. Crowley. World modeling and position estimation for a mobile robot using ultra-sonic ranging. In *Proc. IEEE Int. Conf. Robotics and Automation*, pages 674–681, 1989.

[58] J.L. Crowley. Navigation for an intelligent mobile robot. *IEEE J. Robotics and Automation*, 1:31–41, 1985.

[59] R. Wesson F. Hayes-Roth J.W. Burge C.Statz and C.A. Sunshine. Network structures for distributed situation assesment. *IEEE Trans. Systems Man and Cybernetics*, 11:5–23, 1981.

[60] K.P.Dunn D. Willner, C.B.Chang. Kalman filter algorithm for a multi-sensor system. In *15th IEEE Conf. Decision Control*, Clearwater, Florida, 1976.

[61] P. Dario, D. De Rossi, C Domenici, and R. Francesconi. Ferroelectric polymer tactile sensors with anthropomorphic features. In *Proc. IEEE Int. Conf. Robotics and Automation*, page 332, 1984.

[62] M.H. DeGroot. *Optimal Statistical Decisions*. McGraw-Hill, New York, 1974.

[63] M.H. DeGroot. Reaching a consensus. *J. of the American Statistical Assoc.*, 64(345):118–121, 1974.

[64] D. Djian and P.J. Probert. Case study in formal specification for distributed sensor integration. Technical Report 1844/90, Oxford U. Robotics Research Group, 1990.

[65] B. Donald and J. Jennings. Sensor intepretation and task-directed planning using percetual equivalent classes. In *in Proc. IEEE Intl Conf on Robotics and Automation*, Sacramento, CA, 1991.

[66] M. Drumheller. Mobile robot localization using sonar. *IEEE Trans. Pattern Analysis and Machine Intelligence*, 9(2):325–332, 1987.

[67] H.F. Durrant-Whyte. Consistent integration and propagation of disparate sensor information. *Int. J. Robotics Research*, 6(3):3–24, 1987.

[68] H.F. Durrant-Whyte. *Integration, Coordination, and Control of Multi-Sensor Robot Systems*. Kluwer Academic Press, Boston, MA., 1987.

[69] H.F. Durrant-Whyte. Sensor models and multi-sensor integration. *Int. J. Robotics Research*, 7(6):97–113, 1988.

[70] H.F. Durrant-Whyte. Uncertain geometry in robotics. *IEEE J. Robotics and Automation*, 4(1):23–31, 1988.

[71] H.F. Durrant-Whyte, S. Grime, and H. Hu. A modular, Transputer-based architecture for multi-sensor data fusion. In *Proc. Second Int. Conf. Applications of Transputers*, pages 71–77, 1990.

[72] H.F. Durrant-Whyte, B.Y. Rao, and H. Hu. Toward a fully decentralized architecture for multi-sensor data-fusion. In *Proc. IEEE Int. Conf. Robotics and Automation*, pages 1331–1336, 1990.

[73] A. Elfes. Sonar-based real-world mapping and navigation. *IEEE J. Robotics and Automation*, 3(3):249–265, 1987.

[74] A. Elfes. Integration of sonar and stereo range data using a grid-based representation. In *Proc. IEEE Int. Conf. Robotics and Automation*, page 727, 1988.

[75] O. Faugeras and N. Ayache. Building visual maps by combining noisy stereo measurements. In *Proc. IEEE Int. Conf. Robotics and Automation*, pages 1433–1438, San Francisco, U.S.A., 1986.

[76] O. D. Faugeras, M. Hebert, and E. Pauchon. Segmentation of range data into planar and quadratic patches. In *Int. Conf. Computer Vision and Pattern Recognition*, pages 8–13, 1983.

[77] T.S. Ferguson. *Mathematical Statistics: A Decision Theoretic Approach*. Academic Press, New York, 1967.

[78] M. Fernandez and H.F. Durrant-Whyte. An information-theoretic approach to data-validation. In *American Control Conference*, pages 2351–2355, 1993.

[79] P.C. Fishburn. Subjective expected utility: a review of normative theories. *Theory and Decision*, 13:139–199, 1981.

[80] R.A. Fisher. On the mathematical foundations of theoretical statistics. *Philos. Trans. Royal Society of London*, Sec A(222):309–368, 1922.

[81] W. Fleskes and G. Van Keuk. Adaptive control and tracking with elra phased array radar experimental system. *IEEE Radar-80 Conf. Rec. June*, 1980.

[82] A.M. Flynn. Combining ultra-sonic and infra-red sensors for mobile robot navigation. *Int. J. Robotics Research*, 7(5):5–14, 1988.

[83] M.S. Fox. An organisational view of distributed systems. *IEEE Trans. Systems Man and Cybernetics*, 11:70, 1981.

[84] Y. Gao and H. Durrant-Whyte. A transputer-based sensing network for process plant monitoring. In *Third Int. Conf. Transputers and Applications*. IOS Press, 1991.

[85] T.D. Garvey, J.D. Lawrence, and M.A. Fischler. An inference technique for intergrating knowledge from disparate sources. In *7th Int. Joint Conf. Artificial Intelligence, Vancouver*, pages 319–325, 1981.

[86] M. Gelb. *Applied Optimal Estimation*. MIT Press, 1974.

[87] G. Giralt. Research trends in decisional and multi-sensory aspects of third generational robots. In *Second Int. Symp. Robotics Research*, Kyoto, Japan, 1984. MIT Press.

[88] G. Giralt, R. Chatila, and M. Vaisset. An integrated navigation and motion control system for autonomous multisensory mobile robots. In *First Int. Symp. Robotics Research*, page 191, 1982.

[89] I.J. Good. Twenty-seven principles of rationality. In V.P. Godambe and D.A. Sprott, editors, *Foundations of statistical inference*. Holt, Rhinehart, Winston, 1971.

Bibliography

[90] S. Grime. *Communication in Decentralized Sensing Architectures*. PhD thesis, Oxford University, U.K., 1992.

[91] S. Grime, H.F. Durrant-Whyte, and P. Ho. Communication in decentralized sensing. In *Proc. American Control Conference*, 1992.

[92] W.E.L. Grimson and T. Lozano-Perez. Model-based recognition and localization from sparse range or tactile data. *Int. J. Robotics Research*, 5(3):3–34, Fall 1984.

[93] S. Hackwood, G. Beni, and T.J. Nelson. Torque-sensitive tactile array for robots. In *Robot Sensors Vol. 2, Tactile and Non-Vision*, pages 123–131. Spring-Verlag, 1986.

[94] G. Hager. *Active Reduction of Uncertainty in Multi-Sensor Systems*. PhD thesis, University of Pennsylvania, 1988.

[95] G. Hager. *Task Directed Sensor Fusion and Planning*. Kluwer Academic, Boston MA, 1990.

[96] G. Hager and H.F. Durrant-Whyte. Information and multi-sensor coordination. In *Uncertainty in Artificial Intelligence 2*, pages 381–394. North Holland, 1988.

[97] G. Hager and M. Mintz. Computational methods for task-directed sensor data fusion and sensor planning. *Int. J. Robotics Research*, 10(4):285–313, 1991.

[98] J.C.T. Hallam. *Intelligent Automatic Interpretation of Active Marine Sonar*. PhD thesis, University of Edinburgh, AI Department, 1984.

[99] E. Hanle. Control of a phased array radar for position finding of targets. *IEEE Radar-75 Conf. Rec*, 1975.

[100] S.Y. Harmon. Sensor data fusion through a blackboard. In *Proc. IEEE Int. Conf. Robotics and Automation*, page 1449, 1986.

[101] C.J. Harris. *Application of Artificial Intelligence to Command and Control Systems*. Peter Peregrinus Ltd, 1988.

[102] J. Harsanyi. Cardinal welfare, individual ethics, and interpersonal comparisons of utility. *Journal of Pol. Ec.*, LXIII:309–321, 1955.

[103] H.R. Hashemipour and A.J. Laub. On the suboptimality of a parallel kalman filter. *IEEE Trans. Automatic Control*, 33(2):214–217, 1988.

[104] H.R. Hashemipour, S. Roy, and A.J. Laub. Decentralized structures for parallel kalman filtering. *IEEE Trans. Automatic Control*, 33(1):88–93, 1988.

[105] T. Henderson, W. Fai, and C. Hansen. Mks: A multisensor kernel system. *IEEE Trans. Systems Man and Cybernetics*, 14(5):784–791, 1984.

[106] T. Henderson, E. Weitz, C. Hansen, and A. Michie. Multisensor knowledge systems: Interpreting 3-D structure. *Int. J. Robotics Research*, 7(6):114–137, 1988.

[107] W.D. Hillis. A high-resolution imaging touch sensor. *Int. J. Robotics Research*, 1(2):33–44, 1982.

[108] P. Ho. Organization in distributed sensing. *First year report. Robotics Research Group. Oxford University*, 1991.

[109] Y.C. Ho. Team decision theory and information structures. *Proceedings of the IEEE*, 68:644, 1980.

[110] C.A.R. Hoare. *Communicating Sequential Processes*. Prentice Hall, 1985.

[111] S.A. Hovanessian. *Introduction to Sensor Systems*. Artech House, 1988.

[112] R. Howard. Information value theory. *IEEE Trans. Systems Man and Cybernetics*, 1(2), 1966.

[113] H. Hu and P.J. Probert. Distributed architectures for sensing and control in obstacle avoidance for autonomous vehicles. In *IARP Int. Conf. Multi-Sensor Data Fusion*, 1989.

[114] S. Hutchinson and A. Kak. Multisensor strategies using dempster-shafer belif accumulation. In *Data Fusion In Robotics And Machine Intelligence*. Academic Press, 1992.

[115] K. Ikeuchi and T. Kanade. Automatic generation of object recognition programs. *IEEE Proc*, 76(8):1016–1035, 1988.

[116] S. Iyenger, R. Kashyap, and R. Madan. Distributed sensor networks. *IEEE Trans. Systems Man and Cybernetics*, 21(5), 1991.

[117] H. Jeffreys. *Theory of Probability(3rd Ed)*. Oxford University Press, Oxford, UK, 1961.

[118] G.M. Jenkins and D.G. Watts. *Spectral Analysis and its Applications*. Holden-Day Inc, San Francisco, 1968.

[119] E. Kalai. Nonsymmetric nash solutions and replications of 2-person bargaining. *Int.J.Game.Theory*, 6:129–133, 1977.

[120] R.E. Kalman. A new approach to linear filtering and prediction problems. *Trans. ASME J. Basic Engineering*, 82D:34–35, 1960.

[121] R.L. Keeney and H. Raiffa. *Decisions with multiple objectives: Preferences and value tradeoffs*. John Wiley, New York, 1976.

[122] R.J. Kenefic. Optimum tracking of a maneuvering target in clutter. *IEEE Trans. Automatic Control*, 26(3), 1981.

[123] T. Kerr. Decentralised filtering and redundancy management for multisensor navigation. *IEEE Trans. Aerospace and Electronic Systems*, 23(1):83–119, 1987.

[124] L. Kleeman R. Kuc. Mobile robot sonar for target localization and classification. *Int. J. Robotics Research*, To appear, 1994.

[125] R. Kuc. A spatial sampling criteria for sonar obstacle detection. *IEEE Trans. Pattern Analysis and Machine Intelligence*, 12(7):686–690, 1990.

[126] R. Kuc and B. Barshan. Navigating vehicles through an unstructured environment. In *Proc. IEEE Int. Conf. Robotics and Automation*, pages 1422–1426, 1989.

[127] R. Kuc and M. W. Siegel. Physically based simulation model for acoustic sensor robot navigation. *IEEE Trans. Pattern Analysis and Machine Intelligence*, 9(6):766–778, 1987.

[128] R.M. Kuczewski. Neural network approaches to multi-target tracking. In *IEEE First International Conference of Neural Networks*, 1987.

[129] T. Kurien. Issues in the design of practical multitarget tracking algorithms. In *Multitarget-Multisensor Tracking: Advanced Applications*, pages 43–83. Artech House, 1990.

[130] S. Lee and P. Schenker. Sensor planning with hierarchically distributed perception net. Technical report, Jet Propulsion Laboratory, 1993.

[131] S. Lee, P. Schenker, and J. Park. Sensor-knowledge-command fusion paradigm for man/machine system. In *in Proc SPIE International Symposium on Advanced Intelligent Systems*, Boston, 1990.

[132] E.L. Lehmann. *Testing Statistical Hypothesis 2nd Ed (1st Ed 1959)*. John Wiley, New York, 1985.

[133] B.D. Leon and P.R. Heller. An expert system and simulation approach for sensor management and control in distributed surveillance network. In *SPIE Application of Artificial Intelligence*, volume 786, pages 41–50, 1978.

[134] J.J. Leonard. *Directed Sonar Sensing for Mobile Robot Navigation*. PhD thesis, University of Oxford, 1991.

[135] J.J. Leonard, H.F. Durant-Whyte, and I.J. Cox. Dynamic map-building for an autonomous mobile robot. *Int. J. Robotics Research*, 11(4), 1992.

[136] J.J. Leonard and H.F. Durrant-Whyte. Simultaneous map building and localization for an autonomous mobile robot. In *IEEE Int. Conf. on Intelligent Robot Systems (IROS)*, 1991.

[137] J.J. Leonard and H.F. Durrant-Whyte. *Directed Sonar Navigation*. Kluwer Academic Press, 1992.

[138] V.R. Lesser and D.D. Corkill. Functionally accurate, cooperative distributed systems. *IEEE Trans. Systems Man and Cybernetics*, 11(1):81–96, 1981.

[139] R.D. Luce and H. Raiffa. *Games and Decisions*. John Wiley, 1957.

[140] R. Luo. Data fusion and sensor integration: State-of-the-art 1990s. In Abidi and Gonzalez, editors, *Data Fusion In Robotics And Machine Intelligence*, pages 7–136. Academic Press, 1992.

[141] R.C. Luo and M.G. Kay. Multisensor integration and fusion in intelligent systems. *IEEE Trans. Systems Man and Cybernetics*, 19(5):901–931, 1989.

[142] J. Manyika. A communication system for a decentralized sensor architecture. Master's thesis, University of Oxford, 1990.

[143] J. Manyika and H. Durrant-Whyte. Information as a basis for management and control in decentralized fusion architectures. In *Proc. IEEE Conf. Decision and Control, Tuscon*, 1992.

[144] J. Manyika, S. Grime, and H. Durrant-Whyte. A formaly specified decentralized architecture for multi-sensor data fusion. In *Transputing '91*, pages 609–628. IOS press, 1991.

[145] J. Manyika, I.M. Treherne, and H. Durrant-Whyte. A modular architecture for decentralized sensor data fusion: A sonar-based sensing node. In *IARP 2nd Workshop on Sensor Fusion and Environmental Modeling*, 1991.

[146] J.M. Manyika and H.F. Durrant-Whyte. An information-theoretic approach to management in decentralized data fusion. In *In Proc. Spie 92 Conference. Vol.1828*, 1992.

[147] J.M. Manyika and H.F. Durrant-Whyte. A tracking sonar for vehicle guidance. In *Proc. IEEE Robotics and Automation*, 1993.

[148] J.M. Manyika, S. Grime, and H.F. Durrant-Whyte. A formally specified architecture for decentralised data fusion. In *Transputing '91*, pages 609–628. IOS press, 1991.

[149] P.S. Maybeck. *Stochastic Models, Estimaton and Control, Vol. I*. Academic Press, 1979.

[150] R. McKendall and M. Mintz. Models of sensor noise and optimal algorithms for estimation and quantization in vision systems. Technical Report MC-CIS-87, U. Pennsylvania Dept. Computer Science, 1987.

[151] R. McKendall and M. Mintz. Data fusion techniques using robust statistics. In *Data Fusion In Robotics And Machine Intelligence*, pages 211–244. Academic Press, 1992.

[152] R.J. McKenzie and D.G. Mullens. Expert system control for airborne radar surveillance. *American Institute of Aeronautics and Astronautics*, pages 87–2854, 1987.

[153] P. Medawar. *Pluto's Republic*. Oxford University Press, Oxford, 1984.

[154] A. Mitche and J.K. Aggarwal. Multiple sensor integration through image processing: A review. *Optical Engineering*, 23(2):380, 1986.

[155] H. Moravec. Sensor fusion in certainty grids for mobile robots. In *Sensor Devices and Systems for Robotics*, pages 253–276. Springer-Verlag. Nato ASI Series, 1989.

[156] H.P. Moravec and A. Elfes. High resolution maps from wide angle sonar. In *Proc. IEEE Int. Conf. Robotics and Automation*, page 116, 1985.

[157] C.L. Morefield. Application of 0-1 integer programming to multitarget tracking problems. *IEEE Trans. Automatic Control*, 22(6), 1977.

[158] P.M. Morse and K.U. Ingard. *Theoretical Acoustics*. McGraw-Hill, New York, 1968.

[159] A.G.O. Mutambara and H.F. Durrant-Whyte. Distributed and decentralized robot control. In *American Control Conference*, 1994.

[160] P.J. Nahin. Nctr plus sensor fusion equals iffn or can two plus two equal five. *IEEE Trans. on AES*, 16(3):320–337, 1980.

[161] Y. Nakamura. Geometric fusion:minimizing uncertainty ellipsoid volumes. In Abidi and Gonzalez, editors, *Data Fusion In Robotics And Machine Intelligence*, pages 1–6. Academic Press, 1992.

[162] N. Nandhakumar and J. Aggarwal. Integrating information from thermal and visual images for scene analysis. In *Proc. SPIE Conf. Applications of Artificial Intelligence*, pages 132–142, 1986.

[163] J.F. Nash. The bargaining problem. *Econometrica*, page 155, 1950.

[164] J.M. Nash. Optimal allocation of tracking resources. In *IEEE Int. Conf. Decision and Control*, pages 1177–1180, 1977.

[165] J. Von Neumann and O. Morgenstein. *Theory of Games and Economic Behaviour*. Princeton University Press. 2nd Ed, 1947.

[166] A. Oppenheim and R. Schafer. *Digital Signal Processing*. Prentice-Hall, Englewood Cliffs, NJ, 1975.

[167] N.E. Orlando. An intelligent robotics control scheme. In *American Control Conference*, page 204, 1984.

[168] K. Pahlavan, T. Uhlin, and J.-O. Eklundh. Dynamic fixation. In *Proc. 3rd Int'l Conf. on Computer Vision, Osaka*, pages 412–419, Los Alamitos, CA, 1993. IEEE Computer Society Press.

[169] N. Pal and S. Pal. Entropy, a new definition and its applications. *IEEE Trans. Systems Man and Cybernetics*, 21(5):1260–1270, 1991.

[170] A. Papoulis. *Probability, Random Variables and Stochastic Processes*. McGraw-Hill International Editions, 2nd Ed, 1984.

[171] J. Pearl. *Probabilistic Reasoning in Intelligent Systems: Networks of Plausable Inference*. Morgan Kaufmann Publishers Inc., 1988.

[172] R. Popoli. The sensor management imperative. In Y. Bar-Shalom, editor, *Multi-Target Multi-Sensor Tracking*, pages 325–392. Artech House, 1992.

[173] J. Porrill. Optimal combination and constraints for geometrical sensor data. *Int. J. Robotics Research*, 7(6):66–77, 1988.

[174] G.E. Pugh and D.F. Noble. An information fusion system for wargaming and information warfare applications. In *Decision Science Applications*, volume Rep. 314 AD-A106391, August 1981.

[175] H. Raiffa and R. Schlaifer. *Applied statistical decision theory*. Harvard University Press, 1961.

[176] P. Raiffa. *Decision Analysis*. Addison-Wesley, 1968.

[177] B. Rao, H. Durrant-Whyte, and A. Sheen. A fully decentralized multi-sensor system for tracking and surveillance. *Int. J. Robotics Research*, 12(1):20–45, 1991.

[178] B.S.Y Rao. *Data Fusion Methods for Decentralized Sensing Systems*. PhD thesis, Oxford University, U.K., 1991.

[179] B.S.Y. Rao and H. Durrant-Whyte. A decentralized bayesian algorithm for identification of tracked targets. In *IARP 2nd Workshop on Sensor Fusion and Environmental Modeling*, 1991.

[180] B.S.Y. Rao and H.F. Durrant-Whyte. A decentralized bayesian algorithm for situation assessment. In *Proc. 29th IEEE Conference on Decision and Control*, pages 827–829, 1990.

[181] B.Y.S. Rao, J.M. Manyika, and H.F. Durrant-Whyte. Decentralized algorithms and architecture for tracking and identification. In *In. Vol 2. IEEE Conference on Intelligent Robots and Systems*, 1991.

[182] C.R. Rao. *Linear Statistical Inference and its Applications*. John Wiley, 1965.

[183] J.M. Richardson and K.A. Marsh. Fusion of multi-sensor data. *Int. J. Robotics Research*, 7(6):78–96, 1988.

[184] V. Rohatgi. *An Introduction to probability theory and mathematical statistics*. John Wiley, 1976.

[185] S. Russell and E. Wefald. *Do the right thing*. MIT Press, 1991.

[186] N.R. Sandell, P. Varaiya, M. Athans, and M.G. Safonov. Survey of decentralized control methods for large scale systems. *IEEE Trans. Automatic Control*, 23(2):108–128, 1978.

[187] L.J. Savage. *The Foundations of Statistics*. John Wiley, New York, 1954.

[188] L.L. Scharf. *Statistical Signal Processing*. Addison-Wesley, 1991.

[189] M. Schwartz and L. Shaw. *Signal Processing: Discrete Spectral Analysis, Detection and Extimation*. McGraw-Hill, New York, 1975.

[190] D. Sengupta and R. Iltis. Neural solution to the multitarget tracking problem. *IEEE Trans. Aerospace and Electronic Systems*, 25(1):96–108, 1989.

[191] C. Shannon. A mathematical theory of communication. *Bell Systems Technical Journal*, 27:379–423, 1948.

[192] P. M. Sharkey, D. W. Murray, S. Vandevelde, and I.Reid. A modular head/eye platform for real-time reactive vision. *Mechatronics*, 3(4):517–535, 1993.

[193] H. Simon. Theories of bounded rationality. In C. McGuire and R. Radner, editors, *Decision and Organization*, pages 161–176. North Holland Publishing Co. Amsterdam, 1972.

[194] R.A. Singer and A.J. Kanyuck. Computer control of multiple site correlation. *Automatica*, 7(6):455–463, 1971.

[195] M.I. Skolnik. *Introduction to Radar Systems*. McGraw-Hill, 1980.

[196] R.C. Smith and P. Cheesman. On the representation of spatial uncertainty. *Int. J. Robotics Research*, 5(4):56–68, 1987.

[197] J.L. Speyer. Communication and transmission requirments for a decentralized linear-quadratic-gaussian control problem. *IEEE Trans. Automatic Control*, 24(2):266–269, 1979.

[198] C. Stirling and D. Morrell. Convex bayes decision theory. *IEEE Trans. Systems Man and Cybernetics*, pages 173–182, 1991.

[199] M. Stone. The opinion pool. *The Annals of Statistics*, 32:1339–1342, 1961.

[200] D. Terzopolous. Integrating visual information from multiple sources. In A.P. Pentland, editor, *From Pixels to Predicates*. Ablex Press, 1986.

[201] C. Thorpe. *Vision and Navigation: The CMU NAVLAB*. Kluwer Academic Publishers, 1989.

[202] J.N. Tsitsiklis and M. Athans. On the complexity of decentralised decision-making and detection problems. *IEEE Trans. Automatic Control*, 30(5):440–446, 1985.

[203] E.L. Waltz and J. Llinas. *Multi-Sensor Data Fusion*. Artech House, 1991.

[204] A. Waxman. A visual navigation system. In *Proc. IEEE Int. Conf. Robotics and Automation*, page 1600, 1986.

[205] S. Weerahandi and J.V. Zidek. Multi-bayesian statistical decision theory. *J. Royal Statistical Soc. A*, 144(1):85–93, 1981.

[206] S. Weerahandi and J.V. Zidek. Elements of multi-bayesian decision theory. *The Annals of Statistics*, 11(4):1032–1046, 1983.

[207] L. Weinberg. Scheduling multifunction radar systems. *IEEE Eastcon 77 Record*, 1977.

[208] A.S. Willsky, M.G. Bello, and D.A. Caston. Combining and updating of local estimates and regional maps along sets of one-dimensional tracks. *IEEE Trans. Automatic Control*, 27(4):799–812, 1982.

[209] L.A. Zadeh. A simple view of the dempster-shafer theory of evidence and its implication for the rule of combination. *AI Magazine*, 7(3):85–90, 1986.

[210] J. Zidek. Multi-bayesianity. Technical Report No.05, Dept. of Statistics, University of British Columbia, 1984.

Index

A
action set, 35, 92
action-outcome association, 92, 104
actions and outcomes, 91
amplitude monopulse, 138
architectures, 7, 11
    centralized, 51, 56
    centralized, limitations, 52
    decentralized, 11, 53, 60
    decentralized hardware, 246
    decentralized, features, 53
    distributed, 60
    distributed sensor networks DSN, 11
    hierarchical, 52, 56
    hierarchical, limitations, 53
    modular, 162
    single sensor, 56
Arrow's impossibility theorem, 39, 108
Asymptotic Equipartition Property AEP, 236
autocorrelation
    unbiased, 152
    biased, 152

B
bargaining, 120, 121, 125
base-line, 138, 195
Bayes
    action, 36
    entropy relation, 43
    group action, 37
Bayes Theorem, 24
Bayes, T., 23
Bayesian
    "super", 37
    classification, 82
    decisions, 35
    methods, 227
    min-max, max-min, 39
    paradigm, 23
    robustness, 46
beam pattern, 238
belief theory, 228

C
Central Limit Theorem, 147
classification, 82, 190
    Bayesian, 82
    centralized, 83
    decentralized, 84
    distributed, 84
    from displacement, 190
    hierarchical, 83
    metrics, 206
communication, 34, 48, 89, 117, 226
    decentralized, 61, 85
    hierarchical, 57, 73
    hierarchical classification, 84
    links, 54
    overheads, 54
    hardware links, 246
computation, 116
    costs, 126, 226
conditionality, 95

consistency, 13, 81, 91, 223
control input, 171
convexity, 42, 97
corner, 169
    differential, 240
cost functions, 226
cost-payoff matrices, 10
costs
    computational, 126, 226
    implementation, 226
    temporal, 126
covariance, 26, 66
    innovation, 88
    inverse, 63, 78
    inverse, partial, 77, 79
    matrix, 22, 65, 232
    location, 159, 204, 246
covariance matrix, 11
Cramer-Rao lower bound CRLB, 44, 235
CSP, 230
cueing, 16, 201, 217, 221
cylinder, 169
    differential, 241

D

data, 4
data association, 34, 87, 89, 230
data fusion, 4
    issues, 4, 6
    managed, 6
    methodologies, 7
dead-band, 144, 151
    decision dead-band, 221
dead-lock, 230
decentralized
    architecture, 60
    architecture, 53
    Bayesians, 37
    classification, 84
    decisions, 36, 108
    information filter, 74
    information filter equations, 80

localization, 177
management, 102
management demonstrations, 201
management implementation, 115
non-linear information filter, 78
systems, 11, 13
Decentralized Kalman Filter DKF, 74
decision theory, 10, 35, 93, 126, 131
degree of disregard, 111
Dempster Shafer, 9, 89, 228
detection, 11
differential, 138, 239
    estimator parameters, 151
    feature model, 138
    uncertainty, 144
differential sonar, 15, 138
diffractive targets, 134
direction of propagation α, 238
discrete-time, 64
displacement step, 193, 198, 206, 210
diversity, 3

E

edge, 169
    differential, 241
entropy, 42, 96, 99, 204
    maximum entropy principle, 46
    and Fisher information, 235
    maximum entropy principle, 233
    of normal PDF, 232
environment, 15, 134, 245
estimation, 7, 9, 63
Euclidean space, 21
expectation, 26
expected utility, 36, 94, 225
expert system, 10

Index

Extended Kalman Filter EKF, 69, 166

F
Factorization Theorem, 24, 69
far-field approximation, 237
feature, 244, 245
feature classification
    algorithm, 192
feature models, 167
features
    diffractive, 134
    reflective, 134
field-of-view FOV, 189, 204
Fisher information, 44, 98
    matrix, 44, 67
    and entropy, 235
focus of attention, 15, 141, 162, 224, 230
Fourier transformation, 153
Fraunhofer region, 238
frequency
    effect of, 239
Fresnel region, 238
fully connected, 54, 104
fusion, 4, 28

G
Gauss-Markov, 63
Gaussian, 65, 67, 147
generic decentralized architecture, 246
geometric
    methods, 228
    uncertainty, 22
global
    information, 100
    posterior, 29, 57, 60
    state estimate, 80

H
hand-off, 16, 201, 217, 221
hardware, 221, 243, 246
    limitations, 244
human, 108, 109, 112

I
impulse response, 237
independent likelihood pool, 32, 57, 61, 75, 84
independent opinion pool, 31, 57, 60, 83
inference, 7, 9, 50
information, 42
    Fisher, 44
    matrix, 66
    metrics, 14, 98, 203
    mutual, 43
    Shannon, 42
    source, 20
    state vector, 66
    entropy and Fisher, 235
information filter, 63
    centralized, 72
    decentralized, 74
    decentralized equations, 80
    equations, 71
    hierarchical, 72
    non-linear, 69
    non-linear decentralized, 78
information metrics
    variation with ξ, 210
    variation with FOV, 204
    variation with parameter model, 206
    variation with range, 204
information-theoretic, 14
initialization, 46, 60
innovation, 87, 88, 172
    covariance, 88

interpretation and representation, 6
iterative algorithm, 121

J
Joey, ix, 185
JPDAF, 89

K
Kalman filter, 9, 63, 145

L
Lagrangian, 234
large distributed networks, 226
least squares, 27, 145
likelihood, 191
    function, 23
    information, 100, 101
    vector, 82
likelihood principle, 23
linear opinion pool, 29
lines, 190
live-lock, 230
localization, 165, 169
    accuracy, 177, 180
    decentralized, 177
location covariance, 159, 204, 246
log-likelihood, 43, 66
loss function, 35

M
Mahalanobis distance, 172
managed data fusion, 6
management
    coupled, 128, 203
    decoupled, 130
    imperative, 91
map-building, 15, 165, 179, 200

maximum a posteriori MAP, 26, 83, 85
maximum entropy principle, 46, 233
maximum entropy prior, 234
maximum likelihood ML, 26, 83
mean, 26, 152
minimum mean square error MMSE, 27, 83, 235
minimum variance, 27
mobile robots, 15, 133
    navigation, 165
    sensors, 133
model distribution, 89
models
    statistical model, 147
    probabilistic, 8
    sensor, 132
    physical, 2
models and representation, 8
modularity, 11, 162
moments, 26
monopulse, 138
    amplitude, 138
    phase, 138
motion
    angular, 243
    linear, 243
    relative, 243
multi-sensor systems, 3
mutual information, 43, 107

N
Nash solution, 39, 110
navigation, 15, 165
navigational beacons, 166
nearest neighbour association, 172
Neural networks, 10
neutral behaviour, 42
NNSF, 89
non-fully connected, 54, 55, 81, 226
non-informative prior, 25, 46, 83, 233

Index

normative, 10, 90
normative management
    elements of, 91
    general strategy, 225
    limits of, 224

O

objectivity, 23, 90, 108
observable parameters, 82
observation, 65, 170
    matrix, 65
    model, 21
    noise covariance, 66, 159
    parameter model, 192
obstacle avoidance, 165
occupancy grids, 166
optimal estimator, 25
optimality, 125
orienteering analogy, 137
outcome PDF, 94
OxNav, ix, 185, 194, 203, 231
    group, 185

P

parameter model, 82, 206
parameter set, 82
partial estimates, 77
partial information matrix, 77
partial information state vector, 77
partial posterior
    reconstruction of, 103
perception, 1, 12, 132, 167
personal probabilities, 41
phase monopulse, 15, 138
plane, 167
    differential, 239
planning
    path, 165
    task, 165
point kinematic model, 171
points, 190

Polaroid, 133, 134
posterior
    entropy, 99, 101
    global, 29, 57, 60
    information, 99, 101
    local partial posterior, 61
    local posterior, 57
power spectrum, 153
preference profiles, 41, 97
preferential structure, 93, 94
primary reception, 138
prior
    entropy, 99, 101
    information, 99, 101
    non-informative, 25, 46
    maximum entropy prior, 234
    non-informative, 233
probabilistic fusion, 28
probabilistic information update, 13
probability theory, 8
process noise, 177

Q

qualitative methods, 8
quantitative methods, 7

R

random process, 151
range, 204, 237, 244
rationality, 13, 125
    bounded rationality, 126
receiving threshold, 134
    effect of, 239
redundancy, 3
reflective features, 134
region of constant depth RCD, 135, 166, 172
representation, 6
residuals, 152, 160
risk-averse behaviour, 42, 97

risk-prone behaviour, 42, 97, 98
road map, 18
robot, 1
robust
    algorithms, 229
    systems, 229
robustness, 229
    Bayesian, 46

S

scalability, 12
score function, 44
secondary reception, 138
sensor
    allocation, 10
    classification hand-off, 221
    control, 5
    control parameters, 159
    coordination, 5
    cueing, 217
    hand-off, 217
    inertial, 8
    management, 5, 7
    models, 163
    node, 53
    operational parameters, 147, 161, 203
    planning, 5
    probabilistic model, 158
    synergy, 5
    management, 90, 102
Sensor Communicating Process Architecture SPCA, 230
sensor models, 8, 132
    behaviour and performance, 10
sensor-feature assignments, 201, 211
sensor-target assignments, 10, 104
sensory perception, 1, 133
Shannon information, 42
single sensor system, 56
single-platform-multi-sensor SPMS, 179

singular value decomposition SVD, 22
sonar, 133
    data processing, 137
    differential, 15
    differential sonar, 138
    limitations, 136
    modified devices, 138
    physical model, 134
    return, 140
    Tracking Sonar, 15, 140, 162
    Tracking Sonar, performance, 160, 243
    return, 247
spectral estimation, 151
specularity, 134
state, 96, 102, 113, 129, 169
    estimation metrics, 203
    global estimate, 80
    information state vector, 66
    partial information state vector, 77
    vector, 60, 82
    of nature, 2, 20
state space, 64
state transition, 64, 171
    matrix, 64
    noise covariance, 65
sub-sampling, 227
subjectivity, 24, 90
    subjective prior information, 24, 57
sufficient statistic, 24, 68
survivability, 12
synergy, 5

T

Taylor series, 70, 81
team theory, 227
time-of-flight TOF, 138, 237
Tracking Sonar, 140, 162, 173
    performance, 160, 243
    algorithm, 141

Index

observation, 170
vehicle configuration, 177
transitivity, 95
Transputer, 55, 162, 230, 246
typical set, 236

U

uncertainty, 3
  analysis of measurement, 154
  ellipsoid, 23
  geometric interpretation, 22
  reduction of, 11, 25, 28
  representation of, 8
utility, 11
  axioms of, 41, 94
  comparable utilities, 111, 112
  comparisons, 39, 108, 109
  expected, 36
  function, 35
  group expected utility, 109
  information-based, 94, 224
  non-comparable utilities, 110, 113
  of information, 94
  theory, 40, 90

V

validation, 87, 144, 245
  gates, 87
value of information, 127

W

white process, 153
whiteness tests, 153